A Guide to Analytical Method Development

Practical and Technical Guides for Laboratory-based Chemists

Editor-in-chief:
Michael Walker, *Michael Walker Consulting Ltd, UK*

Series editors:
Lucia Burgio, *Victoria and Albert Museum, UK*
Steven Lancaster, *Domino Printing Sciences, UK*
Diane C. Turner, *Anthias Consulting Ltd, UK*

For a list of titles in this series see: rsc.li/practguides

How to obtain future titles on publication:
A standing order plan is available for this series. A standing order will bring delivery of each new volume immediately on publication.

For further information please contact:
Book Sales Department, Royal Society of Chemistry, Thomas Graham House, Science Park, Milton Road, Cambridge, CB4 0WF, UK
Telephone: +44 (0)1223 420066, Fax: +44 (0)1223 420247
Email: booksales@rsc.org
Visit our website at books.rsc.org

A Guide to Analytical Method Development

By

Victoria Hilborne
University College London, UK
Email: v.hilborne@ucl.ac.uk

Practical and Technical Guides for Laboratory-based Chemists No. 3

Paperback ISBN: 978-1-83916-810-9
PDF ISBN: 978-1-83767-357-5
EPUB ISBN: 978-1-83767-358-2
Print ISSN: 2754-7108
Electronic ISSN: 2754-7116

A catalogue record for this book is available from the British Library

© Victoria Hilborne 2026

All rights reserved

Apart from fair dealing for the purposes of research for non-commercial purposes or for private study, criticism or review, as permitted under the Copyright, Designs and Patents Act 1988 and the Copyright and Related Rights Regulations 2003, this publication may not be reproduced, stored or transmitted, in any form or by any means, without the prior permission in writing of the Royal Society of Chemistry or the copyright owner, or in the case of reproduction in accordance with the terms of licences issued by the Copyright Licensing Agency in the UK, or in accordance with the terms of the licences issued by the appropriate Reproduction Rights Organization outside the UK. Enquiries concerning reproduction outside the terms stated here should be sent to the Royal Society of Chemistry at the address printed on this page.

Whilst this material has been produced with all due care, The Royal Society of Chemistry cannot be held responsible or liable for its accuracy and completeness, nor for any consequences arising from any errors or the use of the information contained in this publication. The publication of advertisements does not constitute any endorsement by The Royal Society of Chemistry or Authors of any products advertised. The views and opinions advanced by contributors do not necessarily reflect those of The Royal Society of Chemistry which shall not be liable for any resulting loss or damage arising as a result of reliance upon this material.

The Royal Society of Chemistry is a charity, registered in England and Wales, Number 207890, and a company incorporated in England by Royal Charter (Registered No. RC000524), registered office: Burlington House, Piccadilly, London W1J 0BA, UK, Telephone: +44 (0)20 7437 8656.

For further information see our website at www.rsc.org

For general enquiries, please contact books@rsc.org

For EU product safety enquiries, please email books@rsc.org or contact Royal Society of Chemistry Worldwide (Germany) GmbH, Römischer Hof, Unter den Linden 10, 10117 Berlin.

Printed in the United Kingdom by CPI Group (UK) Ltd, Croydon, CR0 4YY, UK

Preface

Whether seeking to answer a research question or being presented with a sample of unknown matter, a certain set of fundamental analytical chemistry questions must be raised and answered. This book guides analytical chemists early in their careers, and at various other stages, through the steps in planning and developing appropriate analytical measurement methods: managing the various complexities of sample matrices, selecting techniques for the identification and quantification of analytes and assessing data to streamline the method development process and showing links between method development and method validation. Approaches range from sourcing and using established methods to adaptation and innovation in method development, considering what is appropriate to solving the analytical measurement problem, minimizing cost and optimum use of time.

Knowing where to start when presented with a new chemical measurement problem can be daunting, particularly for a novice analytical chemist. Many established analytical chemists spend much of their careers focusing on a fairly narrow, generic group of chemicals. This book fills a gap in the literature where there are few to no general guides on what to consider when developing chemical measurement methods, considering all aspects and asking the right questions when planning measurement methods so analysts and data users understand what is asked of them and what is needed.

Practical and Technical Guides for Laboratory-based Chemists No. 3
A Guide to Analytical Method Development
By Victoria Hilborne
© Victoria Hilborne 2026
Published by the Royal Society of Chemistry, www.rsc.org

"This book is a great introduction to the book series. It is a must read for anyone and everyone starting or wanting a refresh in their analytical science journey. It sets the scene with definitions and the different aspects to think about and emphasises the importance of project planning all the steps involved to create methods for use, and get the correct results from analytical instrumentation." – *Diane Turner, Senior Consultant and Director, Anthius Consulting Limited.*

"An authoritative call to action to all analytical scientists to consider the impact on environmental sustainability that every day analysis has on the world around us, providing an excellent starting point for those that wish to learn more about improving sustainability in their own laboratories" – *Dr Zoë J. Ayres, President of the Analytical Science Community Council.*

"Chapter 5 provides comprehensive and easy-to-read guidance on analytical method validation and assessment of the performance, in accordance with ISO/IEC 17025:2017 and ISO/IEC 15189:2022. There are appropriate references for readers to further develop their understanding." – *Dr Ruth Hearn, Director at RHearn Ltd.*

Acknowledgements

I extend my sincere thanks to the Royal Society of Chemistry Analytical Science Community Council and the then President Diane Turner for initiating the creation of Practical and Technical Guides for Laboratory-based Chemists, and for asking me to write this introductory book in the series. I also sincerely thank the book proposal and chapter reviewers.

Contents

1	**Introduction**	**1**
	1.1 Introduction	1
	References	9

2	**Sustainability in Analytical Science**	**10**
	2.1 Safety and Sustainability	10
	2.2 Sustainability Systems and Networks	11
	2.3 Sustainable Actions in Method Development	12
	2.4 Safety and Risk Assessment in Method Development	15
	References	17

3	**Principles for Method Development**	**19**
	3.1 Principles for Method Development	19
	3.2 Method Development Decisions – Factors	21
	References	30

4	**Considerations and Strategies**	**31**
	4.1 Considerations and Strategies	31

Practical and Technical Guides for Laboratory-based Chemists No. 3
A Guide to Analytical Method Development
By Victoria Hilborne
© Victoria Hilborne 2026
Published by the Royal Society of Chemistry, www.rsc.org

4.2	Instrument Selection for Chemical Analysis	32
4.3	Detectors	33
4.4	Instrument Selection for Chemical Analysis—Sensitivity and Resolution	36
4.5	Sample Collection	38
4.6	Method Performance Priorities	41
	References	45

5 Validity and Reliability of Analytical Measurement Methods: Data Analytics   47

5.1	Validity and Reliability of Analytical Measurement Methods	47
5.2	Measurement Uncertainty from Sampling	49
5.3	Fitness for Purpose Strategies	51
5.4	How to Validate Methods	52
5.5	Method Performance Characteristics	55
5.6	Measurement Uncertainty	71
	References	74

Subject Index   76

1 Introduction

1.1 Introduction

Analytical science practitioners work with a range of scientific disciplines, their knowledge and expertise helping to solve problems in biosciences, physics, chemistry, engineering, and medicine. They are the gatekeepers for unlocking the burden of proof through measurement by following well established measurement techniques robustly tested by others or by developing new methods. Analytical scientists often work in specialist areas, becoming comfortable in familiar territory, yet they can underestimate how well equipped they are to explore new applications, even those that seem a long way from their particular specialisms. When exploring new chemical measurement routes, it is easy to get lost in a vast number of options, and can be particularly challenging if the composition of a sample matrix, in which the target chemicals reside, is unknown. Whether an analytical toxicologist wants to identify and quantify drugs or poisons, explore levels of polyhalogenated hydrocarbons in soils, or a food analyst wants to know the levels of halofuginone in chicken liver, they will typically deal with complex matrices.

Equipment and techniques for measuring chemicals and chemical properties were often driven by a need to ensure food safety and quality. Horlicks, a manufacturer of malted milk drinks, pioneered the use of a dipping refractometer, a telescope developed by the optical company Carl Zeiss, for dipping into liquid products to view light refraction, defining a refractive index. They created a bespoke analytical technique for a specific, niche application for testing liquid malted milk ingredient concentrations. Further use was forensic investigation of methanol contamination in illicit liquors, which caused many deaths.[1] Later developments of dipping refractometers include fibre optics and temperature compensation for monitoring the quality of a wide variety of liquids.

Portable, rapid, point of interest, or point of care diagnostics in medical establishments, are a high priority in analytical chemical measurement development, with much of the focus being on electrochemical sensors. What may at first seem simple electrochemical quantification of inorganic arsenic in ground water, may be complicated by signal interference from the different chemical analytes and other impurities that simultaneously cause a detector response. This is, however, equally true if using a more sophisticated inductively coupled plasma mass spectrometry (ICP-MS) instrument.[2] An analytical chemist must therefore first consider separating the chemicals of interest before measurement. A certain degree of detector electronics 'noise' or interference maybe acceptable; however, the greatest contributor to signal noise is likely to be the sample matrix. It is important to note that signal noise and interference in any type of detector is not always easy to identify. Systematic uncertainty, a measurement bias caused by unknown variables, is the most challenging regarding detector responses, both qualitatively and quantitatively. A small excess of vitamin C in effervescent tablets, if below the recommended daily intake will not necessarily harm a consumer, but a positive bias in the amount of this ingredient in millions of tablets, will certainly impact profits. Acoustic emission analysis of cracks in a ceramic nanostructure or other surface coatings of deep-sea oil and gas field valves needs high spatial measurement resolution and low signal noise. A crack in any part of the coating will lead to subsequent failure, which may be catastrophic. In contrast, *in situ* monitoring of aquatic and airborne environmental contaminants is likely to have poor spatiotemporal resolution and relatively high measurement uncertainty.[3] Satellite electromagnetic sensing or ground based tracking or grids of *in situ* sensors or instruments in laboratories will have signal noise and measurement uncertainty that are unique to the instrument, the sample, and the chemicals measured. Uncertainty in measurement data therefore accompanies every step of a measurement process. Identifying and quantifying uncertainty is considered as important as identifying and quantifying the chemical analytes of interest. Ishikawa, a fishbone cause and effect diagram, Figure 1.1, is a graphical method to help

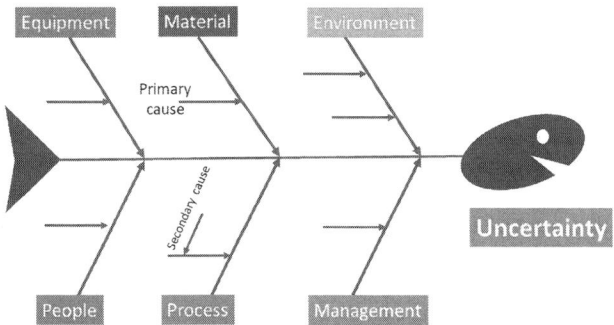

Figure 1.1 Ishikawa – a fishbone cause and effect diagram to map contributions to measurement uncertainty.

identify sources of uncertainty. The impact of various sources of uncertainty on quantification is explored throughout this book and are discussed in detail in Chapter 5, 'Validity and Reliability of Analytical Measurement Methods'.

A wealth of published literature is available to help analytical chemists develop chemical measurement methods, most with peer review qualification, ranging from journal publications, industry and interest group guides, to literature on accreditation such as from the United Kingdom Accreditation Service (UKAS) and Eurachem. The International Organisation for Standardisation/International Electrotechnical Commission (ISO/IEC) provide among others 'General Requirements for the Competence of Testing and Calibration Laboratories' ISO/IEC 17025.[4] Thousands of methods are published every year in multiple journals dedicated to analytical chemistry and the analytical sciences, Journal of Chromatography, Journal of Analytical Chemistry, Talanta, Analytical and Bioanalytical Chemistry, The Analyst, to name a few. Suppliers of chromatography columns and manufacturers of sophisticated benchtop instrument and sensors publish many helpful application notes. Peer review provides high quality literature that helps analytical scientists to perfect their skills through practise and builds banks of knowledge and expertise. The most common approach to analytical measurement method development is adapting and modifying already established methods. Building on prior knowledge is always a wise place to start, why repeat work already done by others? Time spent exploring the literature is time well spent, and can be fascinating, yet requires discipline to maintain focus on the task. Critical evaluation of the literature is essential, asking questions such as: Are the method steps clear enough? Is there sufficient detail or is some information missing?

Are the quoted levels of precision or selectivity realistic for a given sample matrix? By carefully considering the steps for sampling, sample preparation, analysis, data processing and reporting, an analytical chemist can apply and build on prior knowledge.

The aim of this book is to give a general guide for the steps to take when approaching analytical method development, particularly if considering areas that are entirely new to the reader. It will provide support to novices or experienced analysts who want to explore new territory in chemical measurement and help them to evaluate what instruments and sensors an analytical scientist might select for analysing particular combinations of chemical compounds. It will also provide guidance in how to assess their suitability and limitations, alongside associated measurement uncertainty and important data evaluation. Detailed theory on the function and operation of various analytical instruments, their strengths and limitations, are *not* considered. These are instead discussed in the following textbooks of this series, and elsewhere.

1.1.1 Sample Definition

Common sense tells us that poor quality samples and poor planning and preparation will result in poor quality measurement data. A good place to start is to consider the following questions:

1. What is (are) the analyte(s) and *measurand*?
2. What are the samples?
3. What data is needed and is good enough?

The first question 'What is (are) the analyte(s) and *measurand*?' links directly with the third question 'What data is needed and is good enough?' In other words, what information is essential? This guides the amount of effort and expense spent solving a measurement problem. Consider how much time, energy, and other resources should be spent striving for high precision and accuracy (minimum uncertainty), and is the analysis cost justified in terms of the qualitative and quantitative measurement requirements? Equally, time spent getting a new method right will avoid costly mistakes. These three questions therefore link together either linearly, defining the sample, analyte (measurand), measurement system, and data required, or cyclically where improvements are made iteratively. The latter is more likely and is illustrated in Figure 1.2. It is also crucial to recognise whether the data output is fit for purpose. There is no point

Introduction

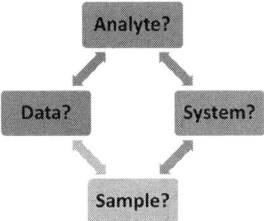

Figure 1.2 Flow diagram of the primary questions an analytical scientist must ask at the beginning of measurement method development.

using anion photoelectron spectroscopy to determine the electronic structure of protein chromophores if the priority is to measure protein biomarkers in waste water to monitor an outbreak of infectious diseases. Is a sophisticated mass spectrometry benchtop instrument measuring femtogram levels of Aflatoxin B1 in grain, corn and nuts necessary if a paper based microfluidic device measuring nanogram levels will suffice? Yet, when developing microfluidic devices, it is important to run confirmatory analysis checks using instruments such as mass spectrometry and nuclear magnetic resonance spectroscopy. Also, for routine checks, at least in the early stages of application, measurements, particularly for health and safety, must be valid and reliable.

To answer question 2, 'What are the samples?', and 3, 'What data is needed and is good enough?', three further questions must also be asked.

1. What is the quickest route to finding what does and does not work?
2. Are you using simplicity over complexity?
3. What is the measurement uncertainty?

Before addressing these questions, it's useful to clarify the definitions of *analyte* and *measurand*.

1.1.2 What is (are) the Analyte(s) and Measurand?

The term for the target chemical is the *measurand*, 'the quantity that is being measured'.

Measurand: *'Knowledge of the kind of quantity, description of the state of the phenomenon, body, or substance carrying the quantity, including any relevant component, and the chemical entities involved.'* JCGM GUM-6:2020.[5]

The *analyte* is the chemical or property that is actually being measured, which may not be the chemical or property of interest as it is not always possible to measure this directly. Both measurand and analyte are often considered as synonyms, however, the term *measurand* explicitly refers to quantity. The conditions used for measurement may change the chemical or physical property of the *measurand*, in which case the quantity actually measured is different from the *measurand* of interest. This emphasises why the term analyte is in common use, which is reasonable considering its noun definition, 'a substance that is analysed or determined, especially quantitatively.' – Oxford English Dictionary.

Consider diffractive imaging showing a variation in cadmium sulphide crystal quantum dot size ranging from 7 to 9 nm. The *measurand* is the size of the cadmium sulphide crystal quantum dots and the *analyte* is the cadmium sulphide crystals. If the toxic peptide linked *N*-acetyl-*p*-benzoquinone imine metabolite of acetaminophen (paracetamol) requires quantification, this is the *measurand*. Then using liquid chromatography with electrospray, triple quadrupole tandem mass spectrometry, the precursor and product ions *m/z* 984.4→535.9 are the *analytes*.[6] To identify and quantify the plant sterol β sitosterol, amongst many plant sterols and stanols, again using liquid chromatography electrospray ionisation with ion trap mass spectrometry, the β sitosterol measurand must be first derivatised using dansyl chloride, hence dansylation. The *measurand* β sitosterol has now become the *analyte* dansylated β sitosterol, or is defined as the new *measurand*.[7] It is always prudent to note that each measurement process step introduces measurement uncertainty. It is unlikely that all of the β sitosterol present in a plant extract sample is derivatised. In other words, if the derivatisation process was repeated, with the same glassware, batches of chemicals, and environment conditions, it is unlikely that exactly the same quantity of β sitosterol is derivatised each time. The process for extracting sterols from plant material is also likely to contribute greater uncertainty in the β sitosterol measurement quantity. This is not necessarily a problem if repeat extraction quantities are consistent. Efficiencies of methods for sample preparation are explored in Chapter 4, 'Considerations and Strategies'.

1.1.3 What are the Samples?

Much of the focus in analytical method development is on the sample preparation, extraction, clean up, analyte recovery and detection sensitivity and selectivity. This is a consequence of historically regarding analytical chemists as solely concerned with the sample once it arrives at the laboratory. Fortunately, current analytical science research and practice increasingly recognises this as a mistake. An analytical chemist must understand the sample, its matrix, to develop effective measurement methods. Quantities and purities of drugs seized by law enforcement agencies indicate a street value that links to a conviction for either possession or supply, possibly resulting in longer prison sentences. Unknown factors from sample matrices that later become known introduce systematic uncertainty, distorting quantification values. Other compounds present in the sample matrix such as impurities and cutting agents can cause suppression or enhancement of mass spectrometry ion detector signal output. Ronidazole is an antiprotozoal agent of the nitroimidazole family and is prohibited by the European Union for food producing animals, *via* the regulation, EC No 470/2009.[8] Isobaric matrix interferences, where another component in the matrix is indistinguishable from the analyte, having near identical mass spectrometry precursor *m/z*, results in false positives for Ronadizole.[9] Non-compliance with the regulations can lead to huge fines and a loss of livelihood. In such high-stake cases, and in general, it is sensible to ensure that there is enough sample to run repeat analysis checks. Is it essential or even possible to run trace analysis? These questions lead to further important questions around what data is needed and is good enough.

1.1.4 What Data is Needed and is Good Enough?

When nuclear magnetic resonance spectroscopy (NMR) was first discovered, the associated early instruments were expensive, cumbersome, and difficult to use. Through an extraordinary feat of collaborative work, commercially available smaller benchtop instruments became affordable. Modern instruments now provide a huge array of sophisticated measurements and some are portable. There is continual development, at an ever-faster pace, of hardware and software for instrument operation and data processing. Keeping up with developments in instrument design, and what is currently commercially available is time well spent. Before selecting an instrument for measuring chemicals of interest, ask questions around exactly what data is essential. Consider whether the equipment, materials, and time spent developing and setting up a method is cost effective while ensuring data reliability and validity. As with the Aflatoxin B1 example and again highlighting the point, why measure the active ingredients in antidandruff shampoos, piroctone olamine and selenium sulphide, to femtogram sensitivity and resolution if typical quantities are a micrograms per 100 ml? The quality control focus must be on appropriate measurement precision (measurement uncertainty), aligning with acceptable batch manufacture precision to ensure product efficacy and safety. Very high sensitivity measurement may be necessary, *e.g.* <1 $mBq\,g^{-1}$ is essential for nuclear forensic investigations and measuring plutonium isotopes in soil environments, from weapons tests or the Chernobyl and Fukushima nuclear accidents.

An environmental analyst may want to simultaneously measure multiple analytes, with quantities of varying magnitude, *e.g.*, types and levels of inorganic carbon, carbon dioxide, bicarbonate, and carbonic acid affecting seawater pH. The environmental analyst might also be an ocean chemist, or work closely with chemical oceanographers to determine which data are important for assessing the impact of atmospheric carbon dioxide solubility on coccolithophore calcium carbonate exoskeletons in seawater. An analytical chemist must work with a range of stakeholders, particularly with those who are commissioning the analysis and so need the data. The measurand data must be continually validated against the measurement system, which means all materials, equipment, and the operator. Quality control systems therefore evaluate the performance characteristics of the measurement method and, by default, the people using them. They evaluate selectivity, repeatability, reproducibility, and measurement accuracy against certified reference materials, bias, ruggedness, uncertainty, and proficiency. These terms and guides to their assessment are peer reviewed, regularly updated and freely shared by Eurachem. Example of Eurachem guides are 'Fitness of Purpose of Analytical Methods',[10] 'Planning and Supporting a Method Validation Study',[11] and 'Use of Uncertainty Estimation in Compliance Assessment'.[12] The validity and reliability of measurement methods should be constantly checked, whether developing new technologies, or techniques and materials, or running routine measurements. This is true whether using sophisticated, benchtop laboratory instruments, satellite, fixed wing, or other mobile instruments in field sampling and measurement, ground based sensors, miniaturised *in situ*, or portable instruments with data logging. A detailed discussion on measurement method validation, techniques for assessing performance and proficiency is given in Chapter 5. When developing analytical measurement systems, the safety, and, increasingly, sustainability are important to consider. Chapter 2 shares ideas for improving sustainability and sustainability innovations for laboratory practice.

References

1. P. P. Parulekar and B. N. Mattoo, *Analyst*, 1978, **103**(1227), 628–631.
2. S. H. Lee, S.-J. Yang, Y. Lee and S.-H. Nam, *J. Anal. Sci. Technol.*, 2021, **12**(1), 28 DOI: 10.1186/s40543-021-00280-8.
3. N. H. Faisal and R. Ahmed, *Meas. Sci. Technol.*, 2011, **22**(12), 125704.
4. ISO/IEC 17025, *General Requirements for the Competence of Testing and Calibration Laboratories*, 2017. www.iso.org [accessed 07-08-2023].
5. JCGM GUM-6:2020, *Guide to the expression of uncertainty in measurement — Part 6: Developing and using measurement models*, 2020. www.bipm.org [accessed 07-08-2023].
6. T. Geib, A. LeBlanc, T. C. Shiao, R. Roy, E. M. Leslie and C. J. Karvellas, *et al.*, *Rapid Commun. Mass Spectrom.*, 2018, **32**(17), 1573–1582.
7. K. N. Franks, L. Alessandroni, G. Caprioli, G. Khamitova, L. Navarini and M. Ricciutelli, *et al.*, *Beverages*, 2021, **7**(3), 61.
8. Regulation (EC) No 470/2009 of the European Parliament and of the Council of 6 May 2009 laying down *Community procedures for the establishment of residue limits of pharmacologically active substances in foodstuffs of animal origin*. http://data.europa.eu/eli/reg/2009/470/oj [accessed 07-08-2023].
9. P. Kumar, A. Rúbies, F. Centrich and R. Companyó, *Meat Sci.*, 2014, **97**(2), 214–219.
10. Planning and Reporting Method Validation Studies, Supplement to Eurachem Guide on the Fitness for Purpose of Analytical Methods, First edition, Eurachem, 2019. https://www.eurachem.org/index.php/publications/guides/planning-validation-studies [accessed 07-08-2023].
11. The Fitness for Purpose of Analytical Methods: *A Laboratory Guide to Method Validation and Related Topics: Second edition*, Eurachem, 2014. https://www.eurachem.org/index.php/publications/guides/mv [accessed 07-08-2023].
12. Use of uncertainty information in compliance assessment, Eurachem, 2nd edn, 2021. https://www.eurachem.org/index.php/publications/guides/uncertcompliance [accessed 07-08-2023].

2 Sustainability in Analytical Science

2.1 Safety and Sustainability

Safety and sustainability are strongly linked in that a positive change in one is likely to be beneficial to the other. Reducing quantities of chemicals inevitably makes laboratory practice safer as well as supporting sustainable use of chemical resources. Examples of sustainable practise in analytical chemistry laboratories include reducing chromatography carrier gas flow rates, switching the carrier gas from helium to hydrogen, reducing the use of hydrocarbon solvents, and using catalysts to reduce heat energy in gas chromatography coupled detectors.[1] Switching from helium to hydrogen may be less safe, hence suitable and sufficient controls must be in place, demonstrating that sustainability can involve challenging choices and compromise. There is increasing interest in improving sustainable practise in chemical measurement laboratories. Analysts are beginning to recognise the negative implications of creating vast quantities of laboratory waste, and that resources are not infinite. It is safer and more cost effective to develop methods using less harmful chemicals, fewer chemicals, less energy, and less water and other consumables. In the not too distant future, as resources dwindle, sustainability risk assessments will become routine. Safety and sustainability implications when analysing only a few samples are quite different to analysing hundreds of thousands of samples, the latter being more

Practical and Technical Guides for Laboratory-based Chemists No. 3
A Guide to Analytical Method Development
By Victoria Hilborne
© Victoria Hilborne 2026
Published by the Royal Society of Chemistry, www.rsc.org

likely to pose serious safety risks as it is likely to involve storing, using, and wasting large quantities of chemicals such as flammable solvents. Laboratory workers routinely write and update safety reviews and risk assessments and it is therefore a logical step to include sustainability risk assessments. Hydrocarbon solvents, gases, other chemicals and disposable plastics all have implications for safety and sustainability. We are sadly all too familiar with the large quantities of disposable plastic waste finding its way into aquatic and land environments, and the negative impacts these have on the inhabitants, including us. Creative thinking for sustainable solutions needs collective encouragement and support in order to develop more sustainable methods for chemical analysis, such as automated microanalysis and with robust data analytics.[2] Miniaturised, robust, easily updated technology and long-lasting equipment, measuring smaller samples, are goals for analytical chemistry research and development. It is also important to consider necessary improvements in working environment conditions, using smaller spaces rather than entire laboratories or buildings. The Sustainable Laboratories Report of the Royal Society of Chemistry, UK, summarises many sustainability challenges and opportunities, noting that starting small is never too small.[3] Analytical chemists are ideally placed to enable sustainable practise, to guide policy, and lead by example in developing and modifying the analytical measurement methods used.

2.2 Sustainability Systems and Networks

A culture of sustainable practice in chemical measurement communities needs fostering. There is substantial progress, which includes several initiatives providing guidance for sustainable practice in science, technology, engineering, and medical laboratories. Initiatives based in the UK are the Laboratory Efficiency Assessment Framework 'LEAF', that provides a wide range of online tools and calculators to guide actions for improving sustainable practice in laboratories:

- Sharing ideas for reducing energy consumption, from ventilation, heating, energy ratings, and the use of ovens, fridges and freezers
- Encouraging judicious use of plastics, other consumables, and water consumption
- Providing workshops and posters to display in laboratories alongside awarding bronze, silver, or gold sustainable laboratory achievements.

A further initiative is the 'Green Gown' award for sharing sustainability lessons and good practice in laboratories across the tertiary education sector, in the UK and internationally. The Green Gown award is sponsored by the UK Research Institute, and awards are given for substantial reduction in institution CO_2 emissions, leadership in creating 'sustainability champion' initiatives and increasing campus biodiversity. Other sustainability networks for STEM subject laboratories, for both industry and academia, are digital platforms such as is 'Warp-it', 'International Equipment Trading Ltd', 'Benchmark' and so on for sharing, recycling and donating equipment, including analytical instrumentation. When communicating with procurement and suppliers of consumables and chemicals, it is important to explore sustainable sources, waste pickup and recycling, and packaging and transport reduction. Analytical scientists use knowledge transfer networks for problem solving and these networks can readily be extended to sharing sustainable practice and sharing what works and what does not. It can be daunting for a novice to join a community of practice networks. Early career analytical scientists are the likely leaders of future sustainable practice, they should be wholeheartedly encouraged in developing their expertise. Innovations often begin through networks, providing the evidence for change, and creating ideas for embedding lifecycle assessments in analytical measurement activities. Successful sustainability initiatives need to shift thinking toward longer term and broader interdisciplinary systems. Legislation, accreditation, certification, and licensing drives much of analytical chemistry practice and using tax incentives, tax on waste, and further legislation should drive improvements in sustainable practice. Public and epidemiological studies ultimately force the setting of legislated limit values and compliance monitoring and, while self-regulation is preferable, it may become increasingly necessary to ensure sustainable use of dwindling resources. Sustainability initiatives led by industry and academia can be an opportunity for guiding government legislation.

2.3 Sustainable Actions in Method Development

Professional organisations, such as the UK Royal Society of Chemistry and commercial industries, fund global sustainability initiatives and research for improving sustainable practice in chemical laboratories. These organisations, among others, support a wide variety of projects, from assessing the viability of storing biological samples in freezers at higher temperatures, washing plastics, and assessing contaminant transfer for reuse, to developing catalysts for more efficient, lower energy demand chemical reactions. An important area of research for improving sustainable practice in laboratories is the reduction in the use and waste of hydrocarbon solvents, finding opportunities to switch hydrocarbon solvents to those that are less harmful in atmospheric chemistry, that do not damage stratospheric ozone, and do not contribute to tropospheric ozone production and further hazards to soil and aquatic environments. Registration, evaluation, authorisation and restriction (REACH)[4] considers phasing out or restricting substances of 'very high concern'. Ideas and actions for sustainable chemical measurement methods are not exhaustive and are likely to change with changing priorities, technologies, and demands. Some common-sense sustainability initiatives include using as few single use plastics as possible, rinsing with water and other solvents, and reusing wherever possible. It is important to keep fume cupboard sashes closed to minimise fan speed, reducing energy consumption while preventing airborne contaminant backflow into the laboratory. Purchase equipment with a high energy efficiency rating. Effective freezer temperature studies show that -80 °C (193 K) is not always essential for storage of biomaterials, -70 °C may be sufficient. The -80 °C recommendation relates to poorer temperature precision and slower cooling rates.[5] As little as a 1 °C increase in biomaterial storage temperature, when applied to many storage freezers over time, can result in a significant reduction in energy use, and hence a reduction in CO_2 emissions. This shows that, collectively, every small action can make a big difference. Further freezer energy efficiency studies explore the application of artificial intelligence – machine learning – to optimise and reduce energy consumption of freezers for commercial food storage.[6] Analytical scientists can use purchasing power to discuss with manufacturers the design of new equipment upgrades. They can also make energy demand and resource efficiency improvements by suggesting functions for shorter run times, reduction in chemical use and efficient shutdown, standby and restart systems.

2.3.1 Negative Results are Positive for Sustainability

Publishing negative results from laboratory synthesis and measurement methods, in open access databases and journals, could substantially help to reduce waste from repeating pointless experiments.

There are a number of journals, such as the American Chemical Society ACS Omega, that could publish negative results without requiring any immediate perceived impact and pharmaceutical companies have internal databases that share 'design of experiment' negative data amongst their employees. Open access databases could be developed to share the strengths and limitations of synthesis or measurement techniques when, for example, switching to more sustainable chemicals, rare elements, solvents, and their sources, or assessing the impact of impurities from recycling. These improvements are not exclusive to laboratory environments, and *in situ* measurements are less likely to be resource intensive than collecting samples and transporting them to a laboratory for analysis. Collected samples are likely to need storage and a variety of preparation, clean up, and preconcentration steps before chemical measurement using multiple resources. Running sophisticated measurement techniques in resource intensive laboratories is likely to be more expensive than *in situ* portable measurements. High quality measurements are always essential for reliable, valid and trustworthy data and the number of in-laboratory measurements will reduce with increasing quality of *in situ* measurements. Democratisation of chemical measurement through portable, *in situ* sample preparation and measurement will also better support the United Nations Sustainable Development Goals (UN-SDGs) and targets. The 17 goals include health equality, clean water and sanitation, environment food security, and sustainable cities.[7] Further to the UN-SDGs are the 12 principles of green chemistry employed some time ago by Galuszka *et al.* (2013), and a crucial principle is 'increased safety'.

When designing sustainable chemical measurement methods, the 12 principles of green analytical chemistry, among other similar approaches to guiding 'green chemistry' practice, are worth consulting.[8] Considering miniaturization of methods, minimising waste, *in situ* direct analysis, eliminating sample pre-treatment, using reagents less hazardous to humans and the environment, minimising energy consumption, and using renewable resources are all of value. Significant advances in electrochemical biosensors, both in portability and miniaturisation, are also of value, through the use of nanomaterials and focusing on real time *in situ* monitoring of redox reactions in the absence of additional reagents.[7] Another example is the use of deep eutectic solvents (DES) as alternatives to hydrocarbon solvents for electrode modification by electrodeposition due to their lower toxicity, renewability, and biodegradability.[9] Other examples of innovations in sustainable chemical measurement techniques, among many, are the exploration of waxed, three-dimensional Origami paper-based enzyme linked assays for the detection of pesticides. Also, edible electrochemical and biosensors derived from charcoal, vegetable oil and vegetable material cellulose with supporting bio catalysis and bio mediated reactions.[9] These innovations may result in substantial changes to established chemical measurement methods, and hence will need strong evidence of the reliability and validity of the measurement data. When establishing, operating or creating new analytical methods, whether these are small or large improvements in sustainability or otherwise, operational safety is paramount.

2.4 Safety and Risk Assessment in Method Development

Every practising analytical scientist must engage with safety and ensure they are fully aware of hazards and hazard controls in each activity, *e.g.* sample collection and storage, sample preparation, analysis, and waste disposal. Every activity must be risk assessed and conform to the Health and Safety at Work Act 1974. Control of Substances Hazardous to Health (COSHH), registration, evaluation, authorisation and restriction (REACH), manual handling, slips, trips, temperature, noise and stress all have important safety considerations. Occupational exposure to chemicals, typically *via* dermal or inhalation exposure, are controlled through European Commission Directives 98/24/EC, 2000/39/EC, 2009/161/EU, and 2004/42/EC. Most analytical scientists are likely to have a good understanding of health and safety at work, with varying degrees of expertise, and of course everyone considers health and safety in their everyday lives outside of the workplace. The UK Institute of Occupational Safety and Health (IOSH)[10] leads on managing safely and safety management training following ISO 45001. A safety risk assessment must accurately describe all activities, chemical use, quantities, and equipment use. It must list all associated hazards then give a clear description of the controls put in place to reduce risk of any serious or harmful consequences from contact with the hazards. This can include a risk matrix where the risk is ranked (green, amber or red) according to the degree of hazard against the effectiveness of the controls, hence the likelihood of severe consequences from encountering the hazards. If the risk matrix score is amber or red, then the hazards and controls must be changed to ensure a green score. There are many parallels

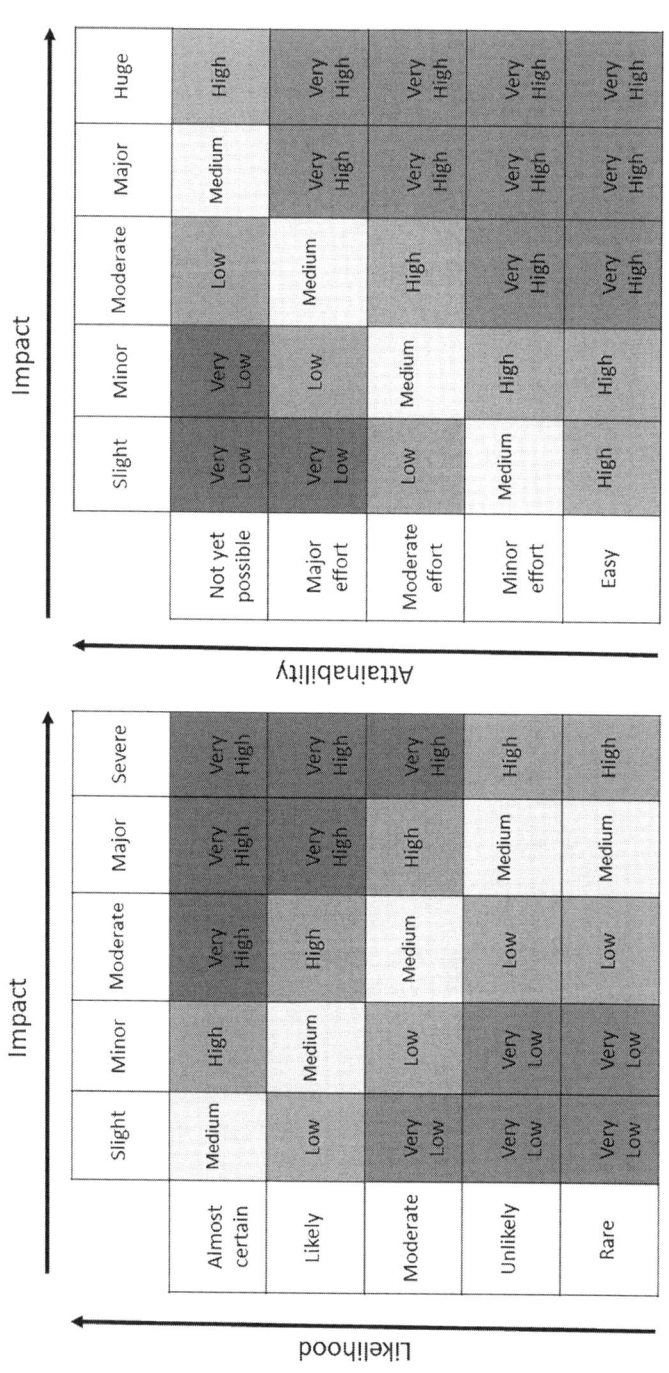

Figure 2.1 Comparison of a safety and a sustainability risk matrix.

between safety risk and sustainability risk assessment and review. The ethos of eliminate, reduce, and ensure safe systems of work apply to both, as illustrated in Figure 2.1. Safety management and workers should routinely review risk assessments and update these upon any changes to activities, hazards or controls. IOSH lists hazard controls in order of their effectiveness for ensuring safety; eliminating a hazard is the most effective control, an example is using spark free refrigerators. Reducing the risk is the next most effective control, for example reducing quantities of hazardous materials, and updating equipment to safer versions and newer technologies. Another control is using barriers, preventing contact with, and between, hazards, flammable substances, and ignition sources. Other controls include using guards on machinery, leak free glove boxes, and ensuring distance from passing traffic. Next, are safe systems of work, ensuring adequate training of operators on the safe handling of materials and equipment and providing clear, accessible instructions detailing how activities should be carried out to minimise risk. Personal protective equipment (PPE) is not generally considered a first-choice control measure, except perhaps in emergencies. For the most part, PPE should address residual risk once the controls are in place. Every person in a workspace is responsible for health and safety and the best health and safety practise is good housekeeping. Being continuously vigilant and monitoring efficient, organised systems of working, using appropriate labelling, keeping good records, appropriate storage, cleanliness, and tidiness are strong approaches to ensuring a safe workplace.

The left matrix in Figure 2.1 is the familiar safety risk matrix of hazard impact and likelihood of the hazard impact occurring in relation to the controls put in place. The comparable sustainability risk matrix plots positive impact on an activity improvement on sustainable practice against the effort needed to make the improvement. Using this tool in many instances takes little effort and focuses on the feasibility so encourages positive action.

References

1. S. Coleman and K. Beard, Agilent Application note, Energy and Chemical, *Analysis of Gas Products from Carbon Dioxide Use Technologies by Gas Chromatography*, Agilent Technologies inc, 2022. https://www.agilent.com/cs/library/applications [accessed 08-08-2023].
2. A. J. Baeumner, G. Gauglitz, L. Mondello, M. C. M. Bondi, S. Szunerits and Q. Wang, *et al.*, *Anal. Bioanal. Chem.*, 2022, **414**(21), 6281–6284.

3. *Sustainable laboratories A community-wide movement toward sustainable laboratory practices*, Cambridge: Royal Society of Chemistry, U.K., 2022. https://www.rsc.org/policy-evidence-campaigns/ [accessed 08-08-2023].
4. Regulation (EC) No 1907/2006 of the European Parliament and of the Council of 18 December 2006 concerning the *Registration, Evaluation, Authorisation and Restriction of Chemicals (REACH)* OJ L 396, 30.12.2006, 1–849 http://data.europa.eu/eli/reg/2006/1907/oj
5. G. Samuel and J. M. Sims, *BMC Med. Ethics*, 2023, **24**, 36.
6. S. Kim, *App. Sci.*, 2023, **13**(1), 346.
7. United Nations Department for Economics and Social Affairs, Sustainable Development, The 17 Goals, 2023. https://sdgs.un.org/goals [accessed 09-08-2023].
8. A. Gałuszka, Z. Migaszewski and J. Namieśnik, *TrAC, Trends Anal. Chem.*, 2013, **50**, 78–84.
9. P. Yáñez-Sedeño, S. Campuzano and J. M. Pingarrón, *Curr. Opin. Green Sustainable Chem.*, 2019, **19**, 1–7.
10. Institute of Occupational Safety and Health (IOSH) Legislation and Compliance, https://iosh.com/guidance-and-resources/business/legislation-and-compliance [accessed 30th April 2024].

3 Principles for Method Development

3.1 Principles for Method Development

The first question is, what do I want to measure and why? A measurement task might at first seem straightforward, but becomes less so when considering whether the measurement data is fit for purpose. Is it necessary to identify and quantify trace amounts of isomers in a manufactured drug to ensure its purity and safety? It is equally important to not over complicate a measurement task. Start by defining what data you need and the types of analysis to get this data. *Qualitative* analysis is the identification of single or multiple analytes in a sample; *quantitative* analysis is the quantification of the measurand (amount of analyte). *Characterisation* determines the physical and chemical properties and/or *fundamental* analytical chemistry, using the underlying theory of analytical chemistry for adapting, improving, and developing a measurement method. Further categories are *screening* for types and groups of chemicals, and *semi-quantitative* analysis of relative and changing quantities in a sample. Screening and semi-quantitative analysis are exploratory rather than targeted analysis.

Both *qualitative* and *quantitative* targeted analysis must compare the measurement data from a sample *measurand* against the measurement data from a reference sample. Referencing should be done against pure standard chemicals of the measurand or the same standards added to a representative sample matrix and whenever possible against a reference material that matches the measurand sample composition. Also, referencing should be done against a blank sample matrix to identify any contamination, and the detector response to the sample matrix without the measurand should be noted. Consider whether the standards are available for purchase from reputable chemical suppliers or, if they are not, can they be made? If there is little to no prior separation of chemicals in a mixture, consider whether any of chemicals present might impact the detector output, either by signal suppression or enhancement, particularly relative to the signal from the pure chemical standard. While mitigation of this problem is possible through good quality sample clean up, preparation, and chromatographic separation, it may not be eliminated entirely. Reactions might occur between multiple measurands and the solvent in a sample, with other chemicals in a sample matrix, or with breakdown products over time, under various environmental conditions, each impacting the detector–analyte signal. More than a decade of studies on calibration of liquid chromatography mass spectrometry using pure standard drug mixtures clearly shows varying degrees of chemical interaction between drug standards in a mixture, hence there is a potential to under or overestimate drug measurand quantities.[1,2] Methods to mitigate and manage this are discussed in Chapter 5.

Increasing sensitivities of chemical detectors mean that *trace* quantitative analysis is commonplace, down to nano and picogram levels of chemicals. There has been recent use of lasers to develop microflow (microfluid) devices that claim to measure attogram levels of antigen–antibody binding.[3] Trace analysis reduces the need to preconcentrate, but increases the need for efficient measurand extraction and clean-up to avoid measurement interference. Trace and non-destructive analysis are very useful when only small amounts of sample are available. This is often the case for forensic investigations, analysing surface swabs for traces of drugs and poisons, accelerants used in arson, and gunshot residue.

Characterisation focuses on the physical and chemical properties or chemical components and materials. Physical properties include particle size and intensity of fluorescence, and chemical properties include structure, reactivity, reaction rates, equilibria, and empirical constants. Examples of empirical constants derived from chemical measurement are diffusion coefficients, important in chromatography and gel electrophoresis, enthalpies of vaporization, pK_a, and equilibrium constants.

Fundamental analysis and analytical chemistry focus on the theories behind chemical separation and measurement. These include phase separation, molecular spectroscopy, rotation, vibration, electronic transition, nuclear spin, ionisation and mass fragmentation, and the chemistry and physics governing analytical measurement instrumentation. Fundamental techniques in analytical chemistry and their application are discussed in detail in the following books of this series. While it is fascinating to know the fundamental principles of chemistry and physics underlying chemical measurement, thorough knowledge and extensive expertise in each or all of these areas is not always essential to develop suitable chemical measurement methods. It is important to recognise that building expertise in developing analytical measurement methods over a wide range of techniques for a wide variety of chemicals, in multiple types of samples and sample matrices, gives a deeper understanding of chemical measurement validity and reliability than fundamental specialism in a few techniques. It is important to build understanding of how to select appropriate chemical measurement techniques, and to process and interpret the accompanying data. Clear definition of data requirements and their purpose is the first step to making good chemical measurement method development decisions, hence ensuring the data is valid and reliable—an important point worth repeating.

3.2 Method Development Decisions – Factors

Consider scenarios A–H listed below, and consider which of these are qualitative, quantitative, characterization, fundamental, screening or semi-quantitative types of analysis.

   A. A waste incinerator is suspected of releasing more than the legal limit of dioxins into the atmosphere.
   B. An art gallery suspects a masterpiece was switched to a forgery.
   C. Airport security needs a more reliable method for detecting explosives in luggage.
   D. Determining the loss of metabolites from a new drug through excretion in urine.
   E. A new law for drug driving needs a method for roadside screening.
   F. Levels of drugs and metabolites in river waters from sewage wastewater treatment plant overflow.

G. A new technique is needed to identify gas clathrates at a deep-sea oil well head.
H. Monitoring for the presence of persistent organic pollutants in food.

The following answers are reasonable: A, quantitative; B, qualitative; C, fundamental; D, characterisation; E, fundamental; F, screening; G, fundamental; H, semi-quantitative.

Analytical chemists know that chemical measurement requirements can fall into more than one of these categories. Qualitative analysis often accompanies quantitative analysis. If samples for measurement are consistent and stable, hence unlikely to change, and if the measurement system is robust, then using fast, cost-effective quantitative methods alone will often suffice. For fundamental analysis, the analytical chemist must first explore data and peer reviewed sources of information to select a suitable measurement technique, drawing from the expertise of others on the optimum methods for measuring specific chemicals in specific chemical matrices. A lack of expertise in the instrument measurement principles means tools for critical analysis of chemical measurement data validity and reliability become ever more crucial to ensure the measurement data is not false and misleading. Also, a lack of expertise in the principles behind chemical measurement techniques and associated instrument hardware means safety advice should be sought to ensure awareness and understanding of potential hazards and associated controls. A good example is laser safety; choosing a portable class 3 laser for quantifying chlorophyll in spinach leaves might cause permanent eye damage if used inappropriately. Figure 3.1. illustrates a class 3 laser housed in plastic casing to block the laser light from the user environment, thus providing a barrier to control the laser hazard, so preventing eye damage.

How appropriate, safe, and sustainable every aspect of a measurement system is, needs thorough review before starting work in a laboratory or in field measurement.

3.2.1 Method Development Decisions

The description of a measurement system must be clear and fully representative. What is the physical and chemical form of the sample? What are the preparation, preconcentration, and separation (if necessary) steps, and what are the methods of detection and data processing? A measurement system considers all inputs and outputs

Principles for Method Development 23

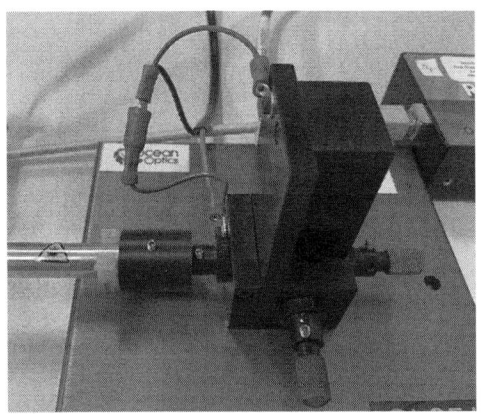

Figure 3.1 Measuring chlorophyl from leaves using a laser with a barrier safety control.

and unless a system is adiabatic, the inputs are likely to include the immediate environment as a sample may be subject to light, temperature, and humidity. Boundaries of concentration range, sensitivity, precision, and selectivity must be established and detection data transformations decided. If investigating the concentration of drug metabolite in a liver sample after ingestion of a drug of specified dose, the transform $y = 0.8x$ indicates an 80% median drug bioavailability (80% of the original drug dose, x, is converted to the active metabolite, y). Repeat extraction of metabolites from hepatic venous blood confirms a median bioavailability of 80%, accounting for a median extraction efficiency of 95%. It is not unusual to achieve extraction efficiencies below 100%. Measurement of drug metabolite concentrations in both the liver and hepatic venous blood are dependent on the efficiency, repeatability, validity, and reliability of metabolite extraction, clean up, and the detector measurement sensitivity. If the metabolite is susceptible to light (photo) degradation, then the system requires detail of further steps for timely prevention of degradation, including data on the efficiency and repeatability of intervention. It is essential to keep accurate records of exact times of sample collection, packaging, and storage, hence continuity, and for each preparation and analysis step. A warning: time can masquerade as an input variable; however, time is rarely a system input. This misconception is not unusual as chemicals can react and degrade over time, but time is not the cause. Consider the sugars in grapes grown for producing wine, the sugar content changes with time, over a growing season and from one year to the next. The input variables effecting this change are sunlight, precipitation, soil nutrients, plant health, and so on. Knowledge of viniculture and keeping accurate records, meta data, serve to guide method development improvements. Another example is whether derivatisation should follow or is integral to extraction to form a more stable measurand. Derivatisation is integral to extraction when quantifying milk sterols which have variable stability under different heating conditions. Understanding which variables are important and which are trivial is typically found through observation, giving essential information for efficient and good quality chemical measurement system decisions.

3.2.2 System Decisions

A system is any physical or chemical process that has measurable parts. The range of types of systems is vast, from molecular separation and cell receptor reactions to supernovae. An analytical pharmacologist or toxicologist measuring drug metabolites, or an environmental chemist measuring persistent organic pollutants in surface water, must define the important input and output variables. Atmospheric carbon dioxide, and bicarbonate and carbonic acid (carbonate) impact sea water pH, as noted in Chapter 1 and in eqn (3.1) below. The system inputs and outputs in this example depend on whether you are considering the forward or the reverse reactions, or both. Calcium carbonate from the exoskeletons of coccolithophores (phytoplankton) could be the input and the pH $[H_3O^+]$ the output of interest if it is important to assess the long-term impact of carbon dioxide absorption on seawater pH, and on limestone formation.

$$CO_2 + H_2O = H_2CO_3 = HCO_3^- + H^+ = CO_3^{2-} + 2H^+. \tag{3.1}$$

It is essential to first distinguish between the factors and the variables. The factor 'dissolved inorganic carbon' has the identifiable and quantifiable variables HCO_3^- and CO_3^{2-}. Unknown factors can affect a measurement system resulting in uncontrolled and erratic outputs. For example, increased ocean acidity increases dissolution of calcium carbonate ($CaCO_3 = Ca^{2+} + CO_3^{2-}$) found in the exoskeletons of coccolithophores. Increased amounts of H^+ (H_3O^+) react with natural CO_3^{2-} to form more HCO_3^- so encouraging further dissolution of $CaCO_3$; this is Le Chatelier's principle. Coccolithophore exoskeletons are an unknown factor that become known, and in ocean environments they are also uncontrolled factors. An analytical chemist must work with an environmental chemist to ensure the integrity of the sample pH, as any adjustment to this will adversely impact the measurement data and its interpretation. Inputs, outputs, the factors, responses, and variables therefore follow *system theory* for chemical measurement method development and associated chemometrics. An analytical chemist is therefore interested in the input factors, the qualitative and quantitative variables, that influence a measurement system. A simple example of input factors could be the types of lamps used in photoionisation detectors (PIDs) for measuring airborne hydrocarbons, and the types and amounts of hydrocarbon vapours to be measured using these PID sensors. The PID sensor output factor is a voltage range, which is specific to each type of PID sensor lamp. The lamp sensitivity hence resolution depends on the type of lamp and type of hydrocarbon vapour being measured. Early stage measurement method development is concerned with which inputs influence a system and which do not.

Consider fermentation in wine production quality control. Inputs can be macro scale such as species of bacteria or yeast, or molecular scale such as amount of adenosine diphosphate and sugars in wine making. The system, illustrated in Figure 3.2, on a macro scale is the yeast species *saccharomyces ellipsoideus*, a qualitative variable of the factor 'yeast' used in the production of alcohol. On the molecular scale, adenosine triphosphate (ATP) synthesis is essential for energy production in yeast cells resulting the metabolism of sugars to alcohols. Alcohol is therefore an output factor in wine making, the primary qualitative output variable is ethanol with the quantitative output variable being the percentage of ethanol produced. There are several other input and output factors with variables involved in wine production systems, including sugars and the elusive output factor

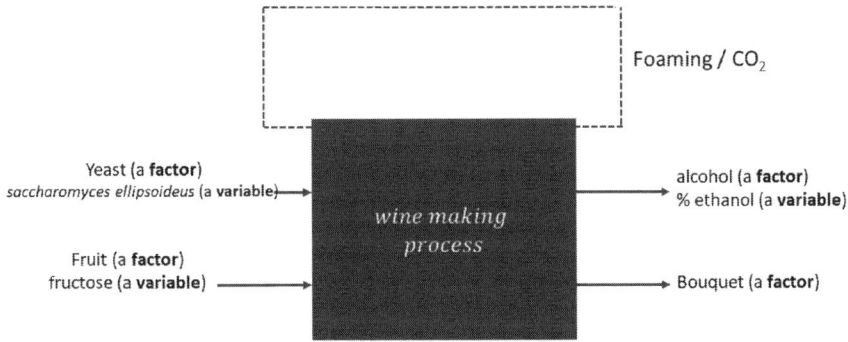

Figure 3.2 Some important input and output factors and variables in wine production.

'bouquet'. Foaming is a masquerading output factor, which results from CO_2 gas production, CO_2 is therefore the true output variable.

A chemical analyst might only be interested in qualitative and quantitative measurement of sugars and alcohols. However, any yeast present in a wine sample would continue the fermentation of sugars present, changing both the sugar and alcohol content over time. This is more likely to be encouraged in some types of lager and beer production rather than wine production. Wine oxidation products could also interfere with the reliability of sugar analysis.[4] The detector response to a chemical variable of interest, ethanol or fructose, as in the wine making example, is the output from an instrument measurement system, the system response.

3.2.3 System Responses

An output factor, or factors, and associated variables are called the system responses. A response surface is an important tool for understanding and optimising chemical measurement system performance. Analytical scientists are familiar with the response surface of a detector signal output from measuring a series of standard concentrations. The detector signal output is likely to have multiple, continuous analogue levels between boundaries of a minimum and maximum signal. However, the ability to quantify step changes depends on the measurement resolution—the detector response to a sample analyte. A response surface is illustrated by plotting the raw, or more commonly the processed, data from the detector response against the system input factor intensive variables of external and internal standard concentrations. It is important to note and check any assumptions that other system factors are constant, typically temperature (isothermal), pressure (isobaric), and volume (isochoric). These should be confirmed by direct measurement and must be true otherwise a response might appear highly variable (noisy) or drift, so introducing systematic error (uncertainty). Not all points on a response surface may be known, if it is reasonable to assume that the input and output data are continuous, then any variable quantity between the minimum and maximum boundaries can theoretically exist. It is therefore possible to estimate any variable quantity between the boundaries using an appropriate model of the relationship between input and output variables. For linear detector signal responses, when voltages and the associated processed data such as peak area and peak area ratio are correlated with a range of concentrations of external and internal standards, the model is a linear regression first order polynomial. While in theory any measurand concentration within minimum to maximum boundaries could result in a correlated response output, as noted above, the limitation is the detector output resolution. All inputs that influence the detector response may not immediately be known and are only discovered after further investigation. Unexpected changes in the chemical form of the measurand can illicit changes in the detector response. An interesting example is aggregation of methylene blue in acidic solution. Methylene blue has multiple uses, which include as a biological stain, an indicator of anionic surfactant contamination, and a redox indicator. Methylene blue aggregation in acidic solutions is seen through small visible absorption spectral shift changes. The associated changes in molar absorptivity, hence deviation from the Beer–Bouguer–Lambert law, depend on the type of aggregation that occurs, which in turn depends on the initial methylene blue and aqueous acid solvent matrix concentration.[5] Aggregation increases over time, introducing systematic uncertainty. The methylene blue aggregate phenomena may not be so easily identified if it is routine to make a fresh sample of methylene blue in hydrochloric acid solution and measure its absorbance immediately. Aggregation is unlikely to be noticeable by sight, particularly with the dark blue of more concentrated solutions where it is more likely for the methylene blue to aggregate until it drops out of solution. To ensure that samples are truly representative, their stability must be considered during each stage of collection, storage, and preparation. Also, it is important to know whether a sample is homogenous or whether to explore any heterogeneity in bulk chemical materials.

3.2.4 Sample Preparation and Collection Techniques

Macroscale samples of chemicals are typically solids, liquids or gases, a further phase is plasma. These may consist of mixtures of different chemicals or a single chemical with single or multiple structural forms. There are thousands of different phases associated with chemical species, which include glassy amorphous solids, crystals, semiconductors, supercritical fluids, gases, and so on. A phase is simply regarded as a uniform state of matter. Analytical chemists usually have to deal with chemical mixtures, and possibly with multiple phases. It is therefore important to know the collection protocols for samples of chemical mixtures, *i.e.* where the samples came from, how they should be stored and how they were transported to the laboratory, if they are not analysed *in situ*. All actions performed on a sample can impact the sample chemical composition, hence the selection of sample preparation and measurement method, and the resultant measurement data. Consider sampling from bulk solids such as from bulk stores of wheat or barley grain, illustrated in Figure 3.3. Is the bulk composition homogenous or heterogenous? Where exactly from the bulk should the grain samples be collected? How many multiples of samples are likely to give an appropriate representation of the bulk? If Ochratoxins, a group of highly toxic and hallucinogenic ergot alkaloid chemicals found in the fungus (the mould) on grains such as wheat and barley, are the measurand, where is it best to collect the grain samples from? Damp environments encourage mould growth increasing the likelihood of finding ochratoxins, particularly if the grain is harvested during a wet season and stored with little or no ventilation. Deciding where to sample from a bulk therefore has important implications regarding food safety and ensuring compliance with food safety legislation. It is likely desirable to homogenise the solid particle size and shape and ultimately the chemical composition of a heterogenous sample. A common method is to mix, cone and quarter for better representation of a mean, median, or mode quantity, and to reduce the sample to a manageable size for preparation and analysis. Agitation of a solid sample with multiple particle sizes will cause size separation under gravity, with the larger particles rising to the top. If a chemical component or measurand of interest is associated with either the larger or smaller particles within a sample, then this will inevitably be concentrated by

Figure 3.3 Filling the grain pile at elevator Creston W.A., Robert Ashworth from Bellingham, WA., USA. Reproduced from ref. 7, https://commons.wikimedia.org/wiki/File:Filling_the_grain_pile_near_elevator_at_Creston,_WA._(36659109360).jpg, under the terms of the CC BY 2.0 license https://creativecommons.org/licenses/by/2.0.

the agitation separation. An analyst therefore will need to back calculate the concentration of the component of interest to a realistic representation of that in the original bulk material. Any concentration values should therefore always include the combined uncertainty from the sampling, sample preparation, and measurement process, giving the overall concentration random uncertainty of precision (repeatability).

When sampling liquids, homogeneity of the measurand in the bulk liquid sample matrix is much more likely than for solids, if the liquids in a sample are miscible. Types and concentrations of chemicals present may change over sample location and time due to changing sources of chemicals in a sample, sample stability, and changing miscibility or solubility. A liquid sample matrix may be highly complex, containing hundreds or thousands of different types of chemicals. Fossil fuel oils contain thousands of aliphatic and aromatic hydrocarbons and associated isomers. Body fluids such as blood or urine contain hydrophilic and lipophilic compounds in blood plasma, red and white blood cell mass, and hydrophilic compounds and proteins in urine. Often, the measurand must be extracted and cleaned before any detection and quantification using instruments or sensors. Gas and vapour samples are most likely to be homogenous; however, as with liquid samples, differences over time and space, including differences in transport, reactivity, and densities, must be considered. Hydrogen gas emitted from radioactive material in water storage is less dense than air, so will rise and collect at the top of closed containers or other enclosed spaces, posing an explosion hazard if there is no hydrogen level control. Increasingly, systems employing direct, in line, gas sampling (introducing the gas directly into an instrument chromatography column and to a detector) use sophisticated timers operating sample valves. Pilot scale assessment for large scale industrial processes might use pulsed sampling and measurement, for example every fifteen minutes over several days. An example is simultaneous analysis of temperature, carbon dioxide, nitric oxide, nitrogen dioxide, and ozone for a study of 'red oil' reaction kinetics using polytetrafluoroethylene tubing and valves for sampling these hazardous gases. Using timer relays to operate the valves and central heating timers to operate the relays precedes work on thermal hazard of tributyl phosphate–nitric acid using microcalorimetry.[6] Strategies for guiding chemical measurement method development, method validation, and data processing protocols, using real case studies, are presented in the following chapters.

References

1. D. Remane, M. R. Meyer, D. K. Wissenbach and H. H. Maurer, *Rapid Commun. Mass Spectrom.*, 2010, **24**(21), 3103–3108.
2. R. Bade, G. Eaglesham, K. M. Shimko and J. Mueller, *Talanta*, 2023, **251**, 123767.
3. T. Iida, S. Hamatani, Y. Takagi, K. Fujiwara, M. Tamura and S. Tokonami, *Commun. Biol.*, 2022, **5**(1), 1053.
4. G. A. Walker, J. Nelson, T. Halligan, M. M. M. Lima, A. Knoesen and R. C. Runnebaum, *Molecules*, 2021, **26**(16), 4748.
5. E. K. Golz and D. A. Vander Griend, *Anal. Chem.*, 2013, **85**(2), 1240–1246.
6. Q. Sun, L. Jiang, L. Gong and J.-H. Sun, *J. Hazard. Mater.*, 2016, **314**, 230–236.
7. Filling the grain pile at elevator Creston W.A., Robert Ashworth from Bellingham, WA., USA, File:Filling the grain pile near elevator at Creston, WA. (36659109360).jpg - Wikimedia Commons CC BY 2.0 https://creativecommons.org/licenses/by/2.0. Flickr by the slow lane at https://flickr.com/photos/90536753@N00/36659109360 (archive). It was reviewed on 23 February 2018 by FlickreviewR 2 and was confirmed to be licensed under the terms of the cc-by-2.0.

4 Considerations and Strategies

4.1 Considerations and Strategies

In this chapter, we explore what to consider when selecting instrument features, including detectors and sensors for chemical measurement, considering what the data requirements are and what strategies help us to meet those requirements. Often method selection is driven by the availability of resources and instrument facilities. Nonetheless, an analytical scientist should maintain awareness of, and consider optimum systems for, measuring chemical and physical properties, particularly in the long term. It is important to ensure data is replicable, so can be shared with fellow analytical scientists and the wider community. In this chapter, strategies for developing chemical measurement systems are reviewed, including:

- Important considerations when selecting instruments for measuring chemicals in sample matrices, and the quality of the data needed.
- Sample collection – what samples are essential, where, when and how to collect these, and the sampling technologies available.
- Sample preparation using common techniques such as liquid and solid phase extraction, micro-extraction, and other less common techniques.
- Strategies for sample storage and stability, both after collection and throughout preparation.

- Comparing and contrasting complimentary strategies for method performance priorities and method performance characteristics. How to tabulate options and the associated data evidence for comparing methods.

The focus of Chapter 5 is method validity and reliability, data interpretation and reporting, and data processing tools for quality assurance.

4.2 Instrument Selection for Chemical Analysis

The physical and chemical properties of a target measurand and associated analytes dictate detector selection for accurate identification and precise quantification. If the detector analyte originates from a complex matrix containing multiple chemicals, then it is better to reduce complexity by extraction, separation, and clean up. If the analyte is present in trace quantities then pre-concentration to a level substantially above the detector limit of detection or quantification is necessary. Whether the measurand is in gas, liquid, solid, or another physical phase will determine the selection of the extraction, clean up, and pre-concentration methods. Many chemical measurement instrument types have sophisticated gas, liquid or solid sample inlet technologies that link to gas or liquid chromatography, then couple to a variety of detectors. In gas chromatography, split/splitless injectors are common and are excellent sample inlet technologies for a vast range of volatile, thermally stable, liquid samples. Programmable, temperature volatilisation (PVT) injectors enhance the vaporisation of less volatile compounds, which is useful when needing to measure multiple components in a mixture with a wide range of volatilities. PTV with multiple column temperature change control and back flushing enables injection onto a gas chromatography column of both volatile and notably less volatile compounds, for example for simultaneously measuring enolones and vanillin derivative extracts from red wine.[1] Once a mixture is injected onto the chromatography column of a suitable stationary phase, chemical separation is optimised by the chromatography method. As the injected chemicals travel in the mobile phase, and preferably not with the mobile phase, they diffuse to varying degrees through the stationary phase, so ensuring separation. Evaluating separation efficiency depends on suitable and sufficient detector signal strength and resolution; a clear, sharp detector signal, substantially above baseline noise, gives narrow, symmetric peaks with measurable baseline separation. Theoretical plates can be used to quantify the separation efficiency and resolution of capillary chromatography columns. This helps with ensuring the chemicals of interest arrive at the detector separately, giving clear, interpretable detector or multiple detector response output. These parameters will of course change with changes in gradient mobile phase.

Details of instrument capabilities, limitations, physics and function, including in-line sample preparation, chromatographic separation and detection are covered in the following books of this series. Check analytical chemistry peer review publications for advances in technology and research, such as advances in chemical synthesis methods controlled by robotics creating a wide variety of new measurands, typically for pharmaceutical and biopharmaceutical development.

4.3 Detectors

Mass spectrometry, nuclear magnetic resonance, Raman, X ray and photon emission spectroscopy detectors are among the optimum for chemical detection selectivity, hence they are among the most reliable for accurate chemical identification. Then, atomic absorption, atomic emission, microwave, infra-red, fluorescence, ultra violet, visible, spectrophotometry, chemiluminescent, thermal desorption spectroscopy, thermal conductivity and electron capture are the next lower level of selectivity or specificity, unless there is little to no matrix interference. Other detection techniques include electronic, acoustic and electrochemical sensing. Following from the discussion in Section 4.2 regarding sample inlets and chromatography, the gas, liquid, solid or otherwise phase of the sample impacts the sensitivity requirements of a detector. Usually there are much lower concentrations of target chemicals in a gas sample at standard ambient temperature and pressure than in liquid or solid samples. However, trace analysis of femtogram levels of a measurand diluted in liquid and solid sample matrices is increasingly common. Ultimately, it is the combination of the physical and chemical properties of a measurand and the associated interaction of analytes with the detector that drives detector selection. Are the target analytes able to absorb and transmit measurable electromagnetic radiation, without interference from other chemicals in the sample matrix, or able to ionise and produce ions that change a detector's electrical response?

Detector responses are usually accompanied by amplification and conversion of an analogue voltage to a digital output signal, it is important that the output signal from a target chemical of interest is as unique as possible, and is selective.

4.3.1 Detector Selectivity

Selectivity is the unique detector signal response to a given target chemical, *i.e.* how discernible the detector signal is from signal responses to any other chemicals present, and from any background signal interference. For clarity, a baseline or background interference signal is typically the detector electronic noise and any detector response to the sample matrix, such as solvents or any other materials present, including unknown contaminants. It is important for an analyst to be alert to the possibility of unknown contaminants being present in a sample, which may result in false positive or false negative detector signal responses.

Using electrospray ionisation with only a single quadrupole mass spectrometer and a detector for identifying caffeine in surface water samples, that are also contaminated with dimethyl phthalate, can result in false positive detector response for caffeine. Positive electrospray ionisation (ESI), using a formic acid modifier, produces precursor ions of 195 m/z for both caffeine and dimethylphthalate, $194.19+1$ m/z for caffeine and $194.18+1$ m/z for dimethylphthalate, so unless the mass spectrometry is high resolution accurate mass, it is likely that false positive detector signals are given if either chemical is unknowingly present. The similarity in molar masses also presents the additional challenge of chromatographic separation. If the separation of these compounds is inadequate, then the presence of dimethyl phthalate is likely to result in an overestimation of caffeine quantification. An alternative is to use a triple quadrupole or ion trap, or a hybrid of the two types of mass spectrometry detector, to produce distinctive product ions from the chemical precursor ions. For precursor and product ions with neutral fragments, multiple reaction monitoring (MRM) gives further identification information and therefore improves analyte identification accuracy. An analyst could consider using high mass accuracy detector technologies to distinguish m/z values to two decimal places; however, these detectors can readily suffer from contamination.

Detector selectivity will always depend on any contaminants present in the sample matrix and the type of target analyte. If sample preparation, extraction, and pre-concentration of target chemicals produces an entirely clean, contaminant free sample, then single quadrupole mass separation with photomultiplier detector response, discernible from the baseline signal, can be enough. Measuring one target chemical at a time is, however, less cost effective than simultaneous analysis of multiple target analytes. The latter is also better for operator efficiency, instrument lifetime, and consumption of energy, solvent, chemical, gas, and other consumables. New technologies to facilitate this are typically initially expensive and require training in their operation, capability, and application, making industry and academia slow to adopt them. Further issues with detector selectivity occur when measuring mixtures of chemicals from a homologous series, isomers or analogues.

4.3.2 Homologous Series Isomers and Analogues

It is usual to simultaneously analyse multiple measurands in chemical groups or with common chemical nomenclature, mycotoxins, pesticides, metals, persistent organic pollutants, atmospheric volatile organic carbons, and so on. When identifying and quantifying groups of chemicals such as these, the selectivity of the detector to each target chemical within the group is likely to be insufficient. Two chemical measurands from a homologous series and their isomers may be separated by chromatography, yet may produce similar, if not identical, detector signals, *i.e.* analyte mass to charge ratios (m/z), hence it is only the arrival time at the detector that differentiates the analytes. Separation of an amphetamine, methamphetamine, methylenedioxymethamphetamine mixture by gas chromatography, and detection with electron ionisation (EI) and single quadrupole mass spectrometry, give too similar mass spectra for clear identification of each of amphetamine. Derivatisation of the amphetamines using perfluorooctanoyl chloride allows for simultaneous detector selectivity and subsequent confidence in quantification.[2] Through derivatisation, the mass spectra for each amphetamine becomes unique giving selective identification. Perfluorooctanoyl chloride is hazardous, so using a less hazardous derivatising chemical agent and alternative detection methods would improve the measurement method sustainability. Amphetamines have the additional challenge of existing as stereoisomers. Advances in chiral separation chromatography with tandem mass spectrometry greatly improves detection selectivity of the different stereoisomers.[3]

4.3.3 Complex Matrices and Selectivity

Many samples are complex matrices, blood, urine, soil, waste, river water, and so on. Extraction, clean up, and concentration of chemicals from complex sample matrices, may still result in fairly complex samples with multiple measurands. These measurands may interact chemically with one another, giving different detector responses from the pure, clean individual target chemical analytes. Ion sources in liquid chromatography tandem mass spectrometry (LC-MSMS) ionisation modes show ergometrine to be most the susceptible ergot alkaloid to matrix effects, negatively impacting its quantification.[4] Ergot alkaloid mycotoxins in rye, wheat, triticale, oat, and barley cereal grains therefore need systematic investigation to find optimum methods of extraction, separation, and matrix matched calibration. Liquid chromatography mass spectrometry signal suppression and enhancement (SSE) matrix effects are commonly found.

The choice of solvent can result in chemical modification impacting a detector response. At pH higher than 1.0 the ^{1}H NMR signals from malic and citric acids overlap, with additional interference from any aspartic acid present in fruit juices.[5] Examination of calibration plots for the analysis of ultra-trace elements in wine, using inductively coupled plasma mass spectrometry, show isobaric matrix interference, ^{54}Cr (^{54}Fe), ^{116}Sn (^{116}Cd), and ^{64}Ni (^{64}Zn).[6] Specificity is therefore synonymous with selectivity, however, application of specificity is often to an entire analytical system including the sample preparation. The specificity and sensitivity of an instrument detector are particularly important for identifying trace amounts of analytes. As the instrument detector response is a function of the method limit of detection of an analyte, then the method limit of quantification of a target measurand is a function of the detector response to the sample matrix. Both are functions of detector resolution.

4.4 Instrument Selection for Chemical Analysis—Sensitivity and Resolution

While sensitivity is important for detecting traces of chemicals, both sensitivity and resolution are crucial for knowing the lowest quantity that an instrument can detect of an analyte. As with selectivity, sensitivity is likely to depend on the analyte of interest and the sample matrix. Scans of fingerprints for traces of drugs, using plasma based dielectric barrier discharge ionisation mass spectrometry (DBDI-MS), have limits of detection for cocaine, heroin and fentanyl at 15, 5, and 54 picogram levels respectively.[7] The presence of the infamous nerve agent Novichok in urine may be measured to a limit of detection (LOD) of 25 $ng\,ml^{-1}$, and degradation product LODs ranging from 10–50 $ng\,ml^{-1}$, using ion chromatography tandem mass spectrometry (IC-MSMS) and ammonium regeneration solution.[8] As measurement resolution is the smallest detectable change in measured quantity, a smaller resolution is likely to also mean a higher measurement sensitivity and possibly precision, but again this depends on the type of analyte and the sample matrix. Continually advancing technologies for measuring very small quantities have led to the design of extremely specialist instrumentation, which eventually become everyday techniques. Radiality analysis of single puncta (RASP), fluorescent bioimaging, shows spatial correlations between microglia, neurons, and α-synuclein oligomers in the human brain. This ultrafine resolution covers relatively wide spatial ranges for complex cellular backgrounds, from single proteins to complex cell phenotypes.[9] To accurately identify and compensate for background signal from unwanted photons and efficiently process large data sets to avoid false positives, are constant challenges. A measurement system may not need the capability for measuring trace quantities, it might just need to be sensitive enough to ensure no false negatives near the detection limit. A detailed discussion on detector resolution, limits of detection and quantification, and their estimation is given in Chapter 5.

4.4.1 *In Situ* Sensing

Monitoring compliance with environmental quality standards (EQS) for priority substances and general pollutants in aquatic ecosystems may use networks of *in situ* sensors.[10] Sensor quantitative sensitivity must be comfortably below the legislated environmental exposure limit. A cartesian grid of *in situ* environment contaminant sensors will have a specified measurement resolution over time and space, depending on the measurement frequency and grid density. Technologies for continuous *in situ* monitoring are increasingly sophisticated, using wireless data communication passing between satellites, to a ground station, then to digital storage, operating in a similar way to mobile phones. Lagrangian tracker type sensors target pollutant transport paths, assuming concurrent contaminant transport with fluid dynamics, measuring concentration fluctuations over time and space. As with Cartesian grids, the measurement resolution depends on measurement frequency. Fixed wing and satellite electromagnetic radiation detection for monitoring atmospheric carbon dioxide or parts per million of particulates from forest fires have incredibly high optical or other electromagnetic spectrum energy detection resolution, particularly considering the distances over which the chemicals of interest are being measured. These produce enormous data sets, requiring substantial data processing and interpretation. Sahara dust and volcanic aerosol measurement using an atmospheric environmental monitoring satellite (AEMS) also called DQ-1 employs a high spectral resolution light detection and ranging (LIDAR) system. The satellite orbits at an altitude of 705 km and measures the vertical profile of aerosols with up to 25% error. At an altitude of around 15 km, the vertical attenuated backscatter coefficient resolution is only 48 m.[13] Other fixed wing and satellite measurement applications include monitoring chlorophyll in forest canopies, phytoplankton in ocean surfaces, the extent of oil spills, ice sheet thickness, to name a few. This is a huge field of analytical science research and practice, where example applications are in environmental forensics and climate change. While these are important to highlight and worth exploring, satellite and large-scale fixed wing measurement data collection and evaluation is somewhat outside of the scope of this book. Methods of sample collection and sample preparation are likely to have the greatest impact on measurement resolution, sensitivity, and precision.

4.5 Sample Collection

Passive and active samplers capture a wide range of chemical contaminants. When using passive or active samplers to capture contaminants in air or water environments, a variety of physical and chemical meta data must be simultaneously collected, measured, and recorded, including temperature, humidity, volume flow rates, position, time, and so on. *In situ* sensing and sampling techniques must therefore consider and compensate for environment complexity.[11] After chemical capture, careful packaging, labelling, and transportation of samples to a laboratory is important. Recording traceability by documenting the sample collection, preparation, extraction, and pre-concentration steps before instrumental analysis is also required.[12] A mass spectrometer detection limit of quantification (LOQ) and resolution, the instrument sensitivity, is less likely to impact quantification limits than the sampling method and extraction efficiency. Environments from which to collect samples of course are not just earth air, soil, and water ecosystems. As highlighted in Chapter 1, systems from which samples are collected are vast, with common systems being blood, bile, digestion, agriculture, food, and drug production. Sustainable methods for chemical measurement mean ever smaller sample sizes, reducing use and waste of consumables for preparation and analysis. Striving for smaller sample sizes rightly raises questions over their representativeness. It is important to ensure that small sample sizes do not result in distorted or misleading inferences about the bulk from where the sample was taken.

As discussed in Chapter 1, Sections 1.1.1, 1.1.2, 1.1.3 and 1.1.4, the first strategy for sample collection is identifying the data needed, and the measurement, accuracy, precision, resolution, and sensitivity requirements. It is also necessary to work closely with users of measurement data to ensure appropriate sampling to ultimately deliver the information that is important to them. Are pulsed, periodic, or continuous sampling methods essential or desirable and are they the most cost effective and sufficient methods for delivering the information needed? Consider blood glucose monitoring where high precision *in situ* continuous glucose monitors are commonplace, yet the output data should still be periodically checked against the traditional measurement of glucose in a finger prick blood sample. Novice users and supporting clinicians should never consider a continuous monitor to be free from systematic or random measurement uncertainty.

Chem-catcher type passive sampler devices for continuous monitoring of contaminants in gas or liquid environments have a wide variety of sophisticated chemical sampling technologies. These methods employ a variety of membrane and chemical sorbents and may embed performance reference compounds (PRCs) to help determine sampling rates (R_s), the volume of sample collection per unit time of these passive samplers. They are able to collect chemicals over extended periods, concentrating trace amounts of chemicals. Examples are polar organic chemical integrative samplers (POCISs) and diffusion gradient thin layer techniques (DGTs). For polar pesticide and pharmaceuticals, the passive samplers consist of types of sorbent materials of wet able polymers, such as styrene divinylbenzene (SDB), C18 and others, sandwiched between poly(ether sulphone) (PES) membranes. For non-polar measurands examples are lipid filled semipermeable membrane devices (SPMDs) and silicone rubber (SR) polydimethylsiloxane (PDMS) or low density polyethylene (LDPE). In relation to LDPE, there is increasing interest in sorption of chemical contaminants onto surfaces of plastic pollutants in natural environments.[14] *In situ* continuous samplers can therefore capture a wide variety of chemicals from lipophilic to hydrophilic, depending on the types of sorbent materials and the permeable membranes used.[15,16] The permeable membranes protect against biofouling of sorbent materials when sampling in aqueous environments, and may control the uptake kinetics. It is wise to be cautious when interpreting the rate of transfer of PRCs to indicate an uptake rate of measurands of interest, particularly when sampling multiple chemicals simultaneously. Many factors can influence uptake rates, not least the water partition coefficient of each chemical of interest, hence the polarity of the chemical measurand, the type of sorbent, physical composition, and membrane–sorbent interaction.[17] These passive sampling devices are increasingly replacing automated programmed control peristaltic pumps and valve grab samplers for sequential sample collection. Whether using bottle rosettes for collecting water samples, storing urine and blood samples in sterile bottles, all are relatively large sample volumes, and are time dependent pulse samples. The data user and analytical chemist must consider the right time and location for sample collection. Spatial sampling for measurement of the thickness of atomic layer deposition of semiconductor metal oxides for gas sensing, using atomic force microscopy, must be consistent across the layer, the 5×5 µm scan areas, with measurements taken at 20 nm intervals.[18] Most measurements are, however, not of such fine resolution. The act of sampling, sample transport, storage, and stability must not alter the measurand of interest in any way. Changes in pressure, temperature, humidity exposure to light, *etc.* may induce chemical reactivity and impact quantity.

4.5.1 Sample Continuity, Transport, Storage, and Stability

Continuity of evidence is essential practice for forensic analysis and in most areas of analytical science application. It is important to run blank sample checks of sampler and sample container contamination. Samples should be sealed in suitable bottles and bags to prevent contamination during transit and storage, using correct and sufficient labelling, and suitably stored, perhaps in fridges or freezers, protecting samples from light sources and thermal or biodegradation. Each of these steps follow strict protocols to ensure sample integrity, continuing with careful recording of steps from opening samples in the laboratory and through the sample preparation and analysis steps. Meticulous recording of all actions taken at each step, often requiring witness signatures to ensure traceability, is required.

Who, when, where and how the collection of samples, storage, transport, preparation, and analysis occurred are used to create an audit trail and are typically recorded digitally *via* laboratory information management systems (LIMSs). Bar codes can be used to support the tracking of samples through a chain of custody, including automated or otherwise analytical sample preparation and measurement processes. Bar code labelling of sample containers, and on sample vials loaded onto and recorded by an instrument autosampler can help to prevent potentially catastrophic and costly mistakes due to sample mix up. For the pharmaceutical sciences, and many other chemistry-based industries, analytical chemists use systems and principles of findable, accessible, interoperable, and reusable (FAIR) data, and they often also use principles of attributable, contemporaneous, original, accurate (ALCOA), where ALCOA+ includes complete, consistent, enduring, and available. Validation steps, as discussed in Chapter 5, ensure that the sampling, sample preparation, and analysis have followed the correct selected measurement method processes and are not subject to contamination at any point. These validation checks are important, not only for assessing and alerting the analyst to incidents of potential contamination, but also for monitoring sample stability, and assessing whether the measurand and subsequent analytes of interest degrade or cross react either between multiple measurands or other chemicals within a sample matrix. Degradation can be initiated by or accelerated by changes in temperature, or exposure to specific light frequencies (photodegradation), changes in pH, in the presence of oxygen, or due to the presence of bacteria enzymes (aerobic or anaerobic biodegradation). If an analyst is unaware of these factors, their measurement data risks being entirely unrepresentative of the initial sample collected at source, and so fails to meet the measurement requirements. As highlighted in earlier chapters, the consequences of this could be costly, dangerous, and even fatal. It is therefore important to run stability tests for guiding analytical measurement method development, and whenever any part of the analytical process is subject to change.

4.6 Method Performance Priorities

Priorities for any analytical measurement method performance are driven by the user requirements of the measurement data. While this statement has been made several times, this is to emphasise its importance as the main driver for developing any analytical measurement method. While it is desirable to have the best quality data possible, this may not always be necessary or cost effective. It is therefore important to clearly state any relevant assumptions, limitations, and warnings accompanying data reporting and presentation. Consider routine measurement for melamine and cyanuric acid in foods, the European Commission Regulation (EU) No 594/2012 states any foodstuff containing 2.5 mg kg^{-1} or more of melamine must be withdrawn and destroyed. This limit is an absolute value, as all analytical measurements have associated random measurement uncertainty (intermediate precision) quantities and should have accompanying checks against measurement bias. It is therefore pertinent to set a sample quantification limit level that is at least the mean plus ten times the quantified random uncertainty of a blank uncontaminated sample matrix from the entire analytical method system, from sample collection, sample preparation to analysis. The laboratory must therefore state with confidence a measurement quantity value at or above the method limit of quantification (MLQ) and below the 2.5 mg kg^{-1} limit. It is better to set a working limit level that is higher than the MLQ, and to check each time whether a MLQ value is applicable when measuring a given sample for its associated analytes. For melamine and cyanuric acid the MLQ is stipulated as being no more than 50 ppb. It is also important to note that a blank intermediate precision, σ, is a population deviation, so each blank sample is from a randomly drawn assumed population of independent, yet identical samples. This is discussed further in Chapter 5.

Although the example 'safe limit' value for melamine and cyanuric acid in food samples is 2.5 mg kg^{-1}, sample sizes collected from a bulk source are likely to be smaller than 1 kg, maybe 100 g, a tenth the size. As discussed in Section 4.5, locations for sample collection, sample quantities, sampling methods, sample containers, and labelling very much depend on a close working relationship with the data user, typically the person or people tasked with the responsibility of ensuring legislation is adhered to. Selection of sampling locations and sampling methods is likely to depend on other specialist knowledge associated with the types of samples and their storage. Several analytical methods are suitable for the analysis of melamine and cyanuric acid in foodstuffs.[19,20] Sample preparation, solid phase or liquid–liquid extraction with liquid chromatography with tandem mass spectrometry analysis offers high level selectivity and quantification sensitivity, yet this is time consuming and expensive. It may be more cost effective to first use reliable screening methods for these measurands and other contaminants of interest before using time

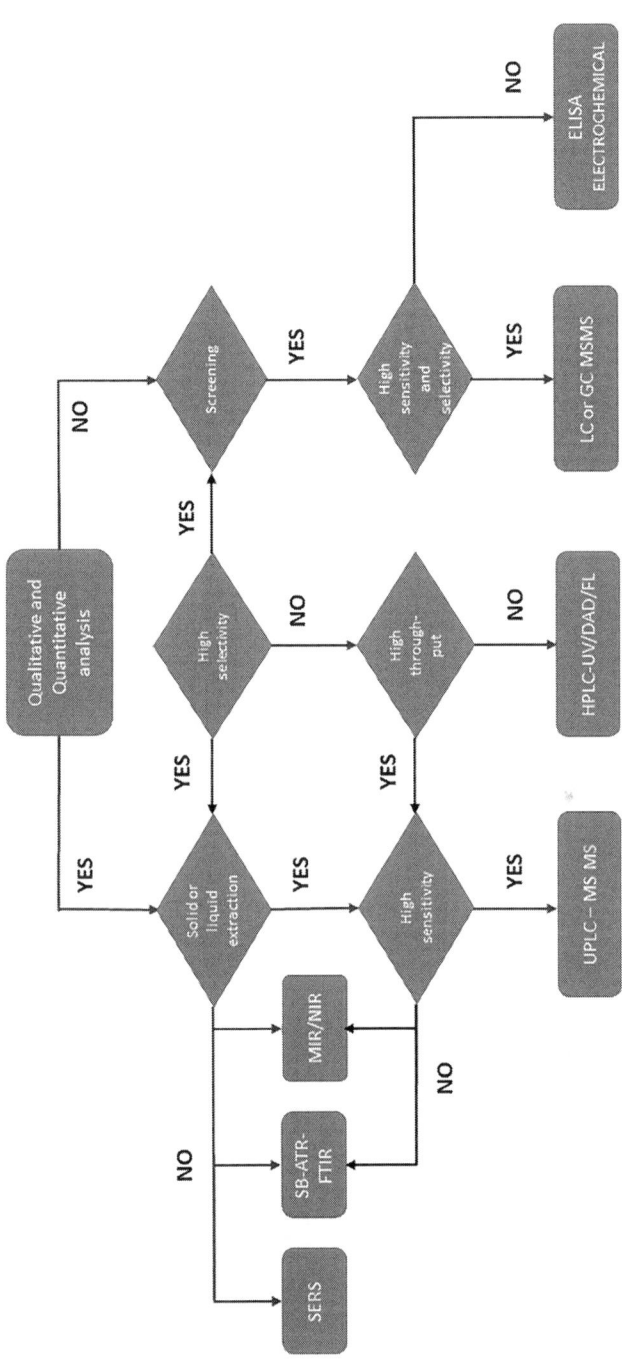

Figure 4.1 Flow chart for example method development considerations: surface enhanced Raman spectroscopy (SERS), single bounce attenuated total reflectance Fourier transform infrared spectroscopy (SB-ATR-FTIR) near infrared (NIR)/mid infrared (MIR), ultrahigh performance liquid chromatography (UHPLC), mass spectrometry (MS), tandem mass spectrometry (MSMS), gas chromatography (GC), enzyme linked immunosorbent assay (ELISA).

Table 4.1 Parameters for consideration when selecting sample preparation and instrumental methods for analysis: solid phase extraction (SPE), liquid–liquid extraction (LLE), solid phase micro extraction (SPME), % relative standard deviation (%RSD), method limit of detection (MLD), method limit of quantification (MLQ). The range is the range of interest for a given measurand or analyte, not the instrument range.

Sample preparation	Matrix	Pre-concentration	Recovery	% RSD	Selectivity
SPE/LLE/SPME	Solid/liquid/gas	Yes/no	>x%	>x%	Yes/no
Quantitative analysis	MLQ	Resolution	SD	Range	Selectivity
Quantity units	>[x]	d signal/d[x]	±δ[x]	Min–max [x]	Yes/no
Qualitative analysis	MLD	Resolution	Interferents	Range	Selectivity
Specificity	>[x]	d signal/d[x]	Yes/no	Nomenclature	Yes/no

consuming and expensive precise and accurate confirmation and quantification techniques. Creation of a flow chart of sample data requirements and measurement method option can be helpful. Figure 4.1 gives a general flow chart of method priorities and commonly used instrumentation for melamine analysis, using only general terms of solid or liquid extraction to cover a wide variety of possible sample preparation methods. A separate flow chart could be created to consider sample preparation methods. There are several other alternative instrumental methods not shown in the flow chart that might also be suitable. It is helpful to tabulate the key parameters of interest, as given in Table 4.1, although these are not exclusive and there are likely to be other parameters of interest, many of which are discussed in Chapter 5, not least those associated with the measurement method data validity, reliability, and uncertainty.

References

1. M. Bueno, J. Zapata, L. Culleré, E. Franco-Luesma, A. de-la-Fuente-Blanco and V. Ferreira, *Molecules*, 2023, **28**, 10.
2. F. Westphal, C. Franzelius, J. Schäfer, H. W. Schütz and G. Rochholz, *Accredit. Qual. Assur.*, 2007, **12**(7), 335–342.
3. B. Kasprzyk-Hordern, V. V. R. Kondakal and D. R. Baker, *J. Chromatogr. A*, 2010, **1217**(27), 4575–4586.
4. S. V. Malysheva, J. D. D. Mavungu, I. Y. Goryacheva and S. De Saeger, *Anal. Bioanal. Chem.*, 2013, **405**, 5595.
5. G. del Campo, I. Berregi, R. Caracena and J. I. Santos, *Anal. Chim. Acta*, 2006, **556**(2), 462–468.
6. M. S. Wheal and E. N. Wilkes, *J. Anal. At. Spectrom.*, 2021, **36**(11), 2383–2390.
7. C. Conway, M. Weber, A. Ferranti, J.-C. Wolf and C. Haisch, *Drug Test Anal.*, 2023, 1–8.
8. M. Otsuka, A. Yamaguchi and H. Miyaguchi, *J. Chromatogr. A*, 2023, **1707**, 464290.
9. B. Fu, E. E. Brock, R. Andrews, J. C. Breiter, R. Tian and C. E. Toomey, *et al.*, *J. Phys. Chem. B*, 2024, **128**(15), 3585–3597.
10. EU. Environmental Quality Standards (EQS) for Priority Substances: Annex I, Part A, Directive 2008/105/EC, 24 December 2008, amended by Directive 2013/39/EU, 24 August 2013 [Accessed 19/08/2024]
11. L. Anhäuser, B. Piorr, M. Arnone, W. Wegscheider and J. Gerding, *J. Exposure Sci. Environ. Epidemiol.*, 2024, **34**(2), 345–355.
12. A. G. Shakallis, H. Fallowfield, K. E. Ross and H. Whiley, *Water*, 2022, **14**(21), 3478.
13. C. Zha, L. Bu, Z. Li, Q. Wang, A. Mubarak and P. Liyanage, *et al.*, *Atmos. Meas. Tech.*, 2024, **17**, 4425.
14. H. Adamu, A. Haruna, Z. U. Zango, Z. N. Garba, S. G. Musa and S. M. Yahaya, *et al.*, *Chemosphere*, 2024, **362**, 142630.
15. N. Reymond, V. Glanzmann, S. Huisman, C. Plagellat, C. Weyermann and N. Estoppey, *Sci. Total Environ.*, 2023, **871**, 161937.
16. A. G. Shakallis, H. Fallowfield, K. E. Ross and H. Whiley, *Water*, 2022, **14**(21), 3478.
17. M. Caban, H. Lis and P. Stepnowski, *Crit. Rev. Anal. Chem.*, 2022, **52**(6), 1386–1407.

18. R. L. Wilson, C. E. Simion, C. S. Blackman, C. J. Carmalt, A. Stanoiu and F. Di Maggio, *et al.*, *Sensors*, 2018, **18**(3), 735.
19. Y.-C. Tyan, M.-H. Yang, S.-B. Jong, C.-K. Wang and J. Shiea, *Anal. Bioanal. Chem.*, 2009, **395**(3), 729–735.
20. A. G. Al-Lafi and I. Al-Naser, *J. Food Compos. Anal.*, 2022, **113**, 104720.

5 Validity and Reliability of Analytical Measurement Methods: Data Analytics

5.1 Validity and Reliability of Analytical Measurement Methods

Validity and reliability of analytical measurements are important for trusting data evidence and for adhering to legislated contaminant limits for consumer products, food, environment, forensic science, medical diagnostics, and so on. International Organisation for Standardisation ISO 15189:2022 – Medical Laboratory Requirement for Quality and Competence, is set 'to promote the welfare of patients and satisfaction of laboratory users through confidence in the quality and competence of medical laboratories'.[1] International Organisation for Standardisation ISO 17025:2017 – General Requirements for the Competence and Testing and Calibration Laboratories, 'enables laboratories to demonstrate that they operate competently and generate valid results, thereby promoting confidence in their work both nationally and around the world'.[2] Use of ISO17025 by many industries facilitates trade by ensuring results are accepted between countries. Validity and reliability of measurements are important in research and development for testing hypotheses and authenticity of discovery and innovation. Several decades ago, Sargent, 1995, wrote four validation in analytical measurement (**VAM**) principles, which still hold true today, and since rewritten as six principles as in the Eurachem Guide – Fitness for Purpose of Analytical Methods, which are listed below.[3,4]

1. Analytical measurements should be made to satisfy an agreed requirement. (*i.e.* to a defined objective.)
2. Analytical measurements should be made using methods and equipment which have been tested to ensure they are fit for purpose.
3. Staff making analytical measurements should be both qualified and competent to undertake the task. (And demonstrate that they can perform the analysis properly.)
4. There should be a regular independent assessment of the technical performance of a laboratory.
5. Analytical measurements made in one location should be consistent with those made elsewhere.
6. Organisations making analytical measurements should have well defined quality control and quality assurance procedures.

Careful and meticulous planning and recording, hence documentation of a procedure, is essential for designing suitable strategies for assessing method validity and reliability. This chapter is a guide to assessing the validity and reliability of measurement methods including data processing, analysis, interpretation, and uncertainty integral to all measurement data. Suggestions for further reading are given throughout. Definitions of qualitative and quantitative uncertainties will depend on the metrics used.

Method validity and reliability are evidenced by the method performance characteristics of selectivity (specificity), calibration and linearity, accuracy (trueness), precision, range, recovery, detection and quantification limits, ruggedness, robustness, fitness for purpose (applicability), and matrix variation. Ruggedness may refer to impact of small changes in method control variables temperature, volume, pH, *etc.* on measurement output data. Whereas robustness might refer to external influences, uncontrolled changes in temperature, other factors, and unknown matrix interferences. Analytical scientists must be aware that clear definition of controlled or uncontrolled parameters is not always possible, and they are unwise to think this is possible as a measurement system is more likely to only have a degree of control. For unknown unknowns such as an unknown matrix component that is causing interference to a detector signal, the impact is on the measurement 'robustness'. Definitions of these performance characteristics are given in the International Union of Pure and Applied Chemistry (IUPAC) Harmonized Guidelines for Single Laboratory Validation.[5] It is important to note that there is no universal agreement on some of the method validation terms, so it is good practice to include summary definitions in the method procedure documentation.

Sampling is the first and arguably one of the most important steps influencing performance characteristics, measurement output data, and associated uncertainty. Discussion around method choices for sampling, sample preparation and analysis techniques are given in the previous chapters.

5.2 Measurement Uncertainty from Sampling

Before running any analytical measurement method, samples must be collected then the samples, not the entire bulk substance, are analysed. An analytical scientist must consider sampling and how this contributes to developing measurement strategies. This is necessary to understand how the data requirements from the bulk material may or may not be met from the samples. The sample or samples may not be homogenous as there may be variation in the amount and stability of chemicals present. Potential matrix interference may influence sample preparation, choice of instrument analysis, data processing and data interpretation. Sampling methods can potentially be the greatest contribution to overall measurement uncertainty. If the choice is to run *in situ* simultaneous sampling and measurement, it is just as important to consider where and when to site sensors and how this will impact measurement uncertainties as it is to define measurement uncertainty from analysis in a laboratory. *In situ* sensors, automated sample collection, preparation and measurement, are common for Earth and other planetary samples, medical diagnoses, and monitoring treatments. An analytical scientist should therefore work with a range of experts to develop sampling strategies to ensure data and any regulatory requirements are sufficiently addressed, and to reduce uncertainty.

5.2.1 Empirical and Modelling Uncertainty in Sampling

Approaches to estimating random uncertainty in sampling are empirical or use modelling. Random uncertainty is the spread or dispersion of repeat, independent, randomly drawn, analyte or measurand quantities from a set of identical samples. See earlier chapters for examples of types of random samples and sampling methods. Empirical random uncertainty is used to determine an overall uncertainty from repeat sampling, ultimately from repeat measurements. Uncertainty modelling is the propagation of individual contributions to uncertainty. This assumes that every contribution is independent from the other and requires knowledge of all contributions. Modelling is useful for evidencing where to make corrections to sampling and measurement methods to reduce random or bias (systematic) uncertainty. Sampling bias—systematic uncertainty, is where the sample collection results in measurement data deviation, or drift, from reference or expected data values and can be difficult to identify. Sampling uncertainty, particularly when sampling from heterogenous and dynamic sources, can be used to confirm sample homogeneity after collection and sample preparation.[6] Table 5.1 lists types of sampling uncertainty. This is a guide as it does not cover all contributions to sampling uncertainty. It is always wise to consult specialist knowledge about the samples.

Further sampling uncertainties, in addition to those listed in Table 5.1, are operator errors in weighing, incorrect assignment of weights or parts of composite samples, or involuntary mixing of sample numbers, and deliberate contamination. Fitness of purpose of sampling methods therefore needs due consideration before any practical work is done.

Table 5.1 Suggested sampling uncertainty types and definitions,[6] recreated from Eurachem Guide Measurement Uncertainty Arising from Sampling: 2019.[7]

Sampling uncertainty type	Definition
Fundamental heterogeneity	Heterogeneity in substance constitution, e.g. chemically or physically different particles.
Grouping and segregation	A result of the distributional heterogeneity in the bulk source.
Long range point selection	Trends over time and space.
Periodic range point selection	Periodic change over time and space.
Increment delimitation and extraction	Correct sample identification. The volume or mass boundaries, shape of a sampling device.
Contamination	Extraneous material, absorption.
Alteration	Chemical: reaction, condensation, precipitation. Physical: disintegration, agglomeration.

5.3 Fitness for Purpose Strategies

The previous chapters introduce a wide variety of measurement methods for a multitude of sample types. While these are wide ranging and novel, they barely scratch the surface of the types and number of analytical measurements carried out across the world daily. Decisions made from these measurements impact livelihoods, safety, ecosystems, and life itself both for the present and future. This seems dramatic, but we often take for granted work done to keep us safe, in servicing our economies, and evidencing policy and decisions for action, so must be 'fit for purpose'. This means the data can be trusted and its reliability and limitations for correct interpretation are known so that decisions can be made with confidence. There is a distinction between verification of data output from running established analytical measurement methods, and validation integral to and following method development work. Any small change, adaptation to the method or method conditions is method development and hence requires validation. A change in the sample matrix of the measurand of interest is likely to need major adaptation to the analytical method despite the resulting analyte being the same. Analytical methods must be reliable and perform near identically under routine conditions, within and between laboratories, with different operators and *in situ* under given conditions. Understandably, it is assumed that the meaning of verification, validity and reliability are known, however there are nuances in documentation giving their definitions.[1,2,8] For ISO17025,[1] verification: *'provision of objective evidence that a given item fulfils specified requirements'*, validation: *'verification where the specified requirements are adequate for an intended use'*. From the Eurachem guide, Fitness of Purpose of Analytical Methods, validation is *'the process of defining an analytical requirement, and confirming that the method under consideration has capabilities consistent with what the application requires'*.[4] Fitness for purpose starts with performance specifications, these include the performance characteristics as listed in the introduction to this chapter plus any extra specified requirements. Ultimately, fitness for purpose is 'good enough' data. On a simple level, does the data output from the method compare to a predefined z score, from an expected population data set, a standard deviation of proficiency testing (SDPT) from multiple laboratories using the same methods?[9] Checking which performance characteristic, if not all, ensures fitness for purpose.

5.4 How to Validate Methods

It is important to carefully document each step in the measurement method procedure, during development and for routine use. Documents should contain a title, introduction, safety, reagents and materials, scope, principle, reactions if applicable, references, including if using published work, sampling, equipment, calculations, proficiency testing (quality assurance and quality control), data presentation, and interpretation. Annexes should be used to cross reference details in method procedures. The scope describes the data requirement, why do the measurement, a list of measurands, analytes (speciation), sample matrix(es), concentrations, concentration ranges, limitations in the method, sample size, sample preparation, and instrument techniques. A summary of the measurement method principle, how it works, including calibration, blank correction, derivatisation, and calculations should also be given. Once satisfied with the within-laboratory performance characteristics, further validation *via* interlaboratory comparison (collaborative studies) might be considered.[10,11] When developing a new measurement method, evaluation of all performance characteristics is sensible. Once the method is established, selection of performance characteristics for validation will depend on the requirements, risks, and resources available. A risk assessment on omission of any method validation tests, considering the impact if any validation tests are not used routinely or intermittently, should be carried out.

5.4.1 Validation Samples

This section summarises important samples for validation planning and reporting, potential pitfalls, and tips. It does not give a detailed, step by step, guide for setting a validation plan as any plan will always be unique to the measurement and associated data requirements. Refer to the documentation referenced at the end of this chapter for further details on validation planning. Acceptance criteria, the validation tools listed in Table 5.2, for each performance characteristic, need defining and documenting.

All types of blanks listed, not least the sample blank hence procedural blank, are important for ensuring that a positive detection signal is attributable to the measurand(s), the detector analyte(s) of interest. An analyst must be vigilant in ensuring that qualitative data is not a false positive or risks a false negative. Procedural blanks reflect some, if not all, of the necessary sample preparation steps prior

Table 5.2 Method validation tools.

Tool	Validation
Sample blank	Sample matrices without any measurand(s) (detector analyte(s)) are present or are below detectable levels.
Procedural blank	A sample blank that has gone through the entire analytical method, sample filtering, extraction, preconcentration, derivatisation, and so on, with the aim of reducing the complexity of the initial matrix to a simpler matrix of, for example, solvents and modifiers.
Calibration blank	A calibration standard minus the chemical measurand (detector analyte) that is used for calibration standards. This is either (i) the procedural blank for matrix matched calibration or (ii) a selected matrix, e.g. solvent, modifier, internal standard.
Reagent and solvent blanks	A mix of the solvents and reagents present in a sample put into an instrument for analysis. This is likely also a calibration blank.
Routine test sample (real sample within accepted stability criteria, having undergone validated measurement)	Used as a reference material for measurement precision and bias checks. This may be a suitably qualified certified reference material (CRM).[12]
Spiking sample matrices	Deliberate addition of the measurand (chemicals of interest or chemically similar or interferents) to a sample matrix, either to a sample blank or with some measurand already present, as in standard addition. For the latter, ensuring that the total concentration is within the working range, or preferably, within calibration range.
Incurred materials	Referring to the bulk sample where something unexpected and unwanted (contaminant) has incurred into the bulk, homogenously or heterogeneously. Pesticides or food allergens are examples, among many.
Measurement standards	Measurement standards often mean guide documents such as ISO standards and statutory instruments. Also, samples containing pure chemicals, or reference properties for identification confirmation and calibration.

to instrument analysis. Any detector signal from a sample blank, hence procedural blank, relative to the calibration, and reagent and solvent blanks, can offer important information on the impact of chemicals other than the target analytes in a sample matrix. Impacts such as changes to the measurand and/or analyte chemical composition, detector signal enhancement or suppression, are examples discussed in previous chapters. Spiking with a known concentration of the measurand (chemical of interest) is a useful method validation and control tool, indicating changes in measurement precision or bias (drift). Caution must be taken when spiking matrices with the chemicals of interest as these may not have the identical form as in the real sample. In a real sample, the chemical of interest may be bound to a protein or bonded in a complex with other chemicals after exposure to physical and chemical processes. Reference materials, or certified reference materials (CRMs) offer similar precision and bias checks. It is important to ensure there are no changes to the measurand (chemical of interest) due to CRM sample instability. As a reference material is a sample matrix containing real amounts of the components of interest, and while it may not be certified, its measurement by competent analysts can still yield useful validation information for method development. Checks on detector signal influence, so data output, from a laboratory reagent or another material are always sensible. It is important to note that CRMs are strictly controlled, having documentation of measurement uncertainty, metrological traceability, and stability.[12]

5.4.2 Measurement Validation Planning

Strategies for measurement validation planning and processes must focus on customer needs, the problems that need solving, and hence produce data that is valid and reliable. Is the measurement method already established in the laboratory, or do simple, minor adjustments to the method need to be made? Must established methods be followed, and regulations and limit values referenced? Analytical requirements should be matched with performance characteristics and examples of applications such as polyfluorinated alkyl substances in foods and drinking water are discussed in previous chapters. It is important to not be concerned if requirements need substantial method development as this may give opportunities for creating more sustainable, efficient and cost-effective methods. Customers often state what methods data processing and reporting they want. There are however, increasing numbers of small medium enterprises, and associated customers who are looking to analytical scientists for measurement advice. Good communication, ultimately, good listening skills, ensures an analytical scientist uses measurement methods that are 'good enough' for customer needs, that are fit for purpose, and clearly agreeing the choice of validation checks and any limitations is essential. Planning starts with sampling and the uncertainty in measurement data attributable to sampling, as discussed in Section 5.2. Whether the measurement laboratory does or does not collect samples, all meta data, time, location, type, conditions, sampling methods, and so on should accompany each sample. Similar meta data should also be recorded if simultaneously sampling and measuring *in situ*. Sample documentation must be scrutinised to identify any interaction with the sample that might cause measurement bias or compromise sample integrity—see earlier discussions giving examples. All sample preparation, homogenisation, subsampling sample size, containers, and storage must also be recorded. Sub samples taken from a bulk can be known as the 'test sample'. Additional sampling procedures can subsequently impact choices for further sample preparation and measurement. Consideration should be given to calibration ranges, instrument selection for detection and quantification limits, random uncertainty (precision), and spiking concentrations. The careful replication of all sampling, sample preparation, and analysis steps should be documented to ensure precision data is reliable, minimising measurement uncertainty from the process and analyst. Consider using robustness or ruggedness tests as part of the validation plan to verify impact from temperature change and other variables influencing sample stability and integrity. An important validation check is comparing measurement data on the same sample using two different techniques. Common practice is use of confirmation checks to compare in field, *in situ*, sampling and measurement with sample preparation and analysis in the laboratory. Care must be taken as measurement methods are often far from identical, hence validation checks for both methods are essential for reliable comparison.

5.5 Method Performance Characteristics

An important first step in method validation is ensuring, when selecting instruments for chemical measurement, that the detector signal is selective to the analyte of interest, *i.e.* that it is discernible from potentially a vast number of other chemicals that might be present in the sample. It should be ensured that the analyte of interest is detectable at the levels and quantities of interest, that the detector is sensitive enough and able to measure the degree of change in quantities as needed. Measurement uncertainty quantities are not always formally considered a performance characteristic. This is a mistake, as uncertainty, particularly measurement data dispersion, random uncertainty, is an inextricable component of the measurement data. Precision and similar dispersion parameters, are unlikely to be fully tested routinely due to analysis costs. Instead, the assumption that a pre-determined measurement precision applies to all subsequent measurements should be used, following the central limit theorem. This assumes that each sample measurement is a randomly drawn replicate from the same population, hence population distribution. This may be reasonable, to a point, for many types of sample measurements, but predetermined precision and the associated assumptions need checking periodically.

5.5.1 Calibration

Instrument calibration is fundamental to chemical and physical property measurement. Calibration is determining the regression parameters of a detector response to an amount of analyte, eqn (5.1). While a linear regression correlation is common, the detector response to an amount of analyte might follow a higher order polynomial correlation, such as logarithmic. If a detector response is non-linear then various software algorithms may be used to linearize the response output. The calibration plot is typically the intrinsic values of concentration, the independent variable, on the x axis, and the detector signal, dependent variable, on the y axis. The linear regression parameter model constants of gradient, m, and intercept, c, eqn (5.1), are illustrated in Figure 5.1.

$$y = mx + c, \text{ including uncertainties, } (y \pm \delta y) = (m \pm \delta m)(x \pm \delta x) + (c \pm \delta c) \quad (5.1)$$

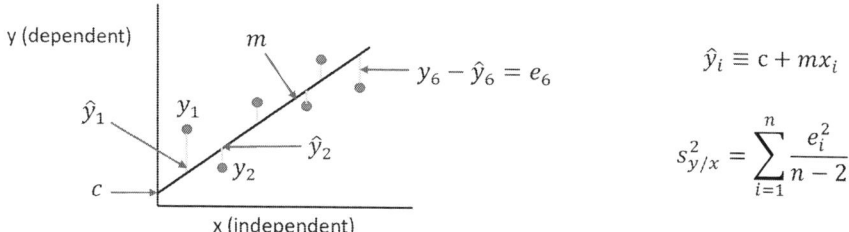

Figure 5.1 Ordinary least squares linear regression fitting. An uncertainty in the intercept greater than the intercept parameter value indicates the calibration regression fails at low concentrations. $R^2 = \pm 1$ indicates a linear correlation between the x and y variables. High residual, e_i, uncertainties show a higher gradient uncertainty, and the residuals are a better indication of linearity, goodness of fit, than a Pearson co-efficient of determination, R^2 value.

where y is the detector signal output, x is the concentration of analyte or measurand of interest, m is the calibration linear regression gradient, and c is the intercept value of y when $x = 0$. The associated uncertainty in each of these values, δ, are important to define during method development. $\hat{y}$ is the estimate of y using the linear regression model parameters of slope and intercept. $s_{y/x}^2$ is the variance in the y residuals given (conditional on) x. Good calibration precision when routinely performing a calibration is a crucial aspect of a method validation or verification study for quantitative analysis. Calibration of instrument detection focusses on the instrument selectivity as discussed above. Figure 5.1. illustrates an ordinary least squares calibration linear regression. A total least square regression includes deviation, bias, in the individual (mean or median) calibration concentration values from the model concentration regression values, $x_i - \hat{x}_i$, which should be as close to zero as possible. The bias in x is therefore any difference from the linear regression model, $\hat{x}$. Random and/or systematic uncertainty in concentration, δx, should be no more than a fifth the magnitude of uncertainty in detector signal output. Ultimately any bias in a single measurement should be within the expected random uncertainty derived from repeat measurements and should not have a consistent bias direction. It is important to note that each calibration plot represents a specific calibration done at a specific time using a specific instrument. Ultimately, better representation of these residuals is from repeat calibration measurement data over a longer period in time, ensuring that there is no bias, drift in signal response over time. The optimum is for any bias to be small, close to zero, hence close to a 'correct' model. Propagation of relative uncertainties in mass and volume are included in the modelling method for determining uncertainty in concentration, x, values. To determine an uncertainty in signal output, δy, at each calibration concentration, at least six replicates should be run. Often, only one calibration concentration is routinely measured in triplicate for each calibration run. Calibration reproducibility, hence long-term precision on different days, by different analysts, are not routinely checked. Calibration is instead validated against measurements of spiked sample recoveries and certified reference materials (CRMs), first checking that there are no issues with these samples, such as any changes in concentration due to degradation introducing measurement bias. Figure 5.2 illustrates a linear regression calibration plot including random uncertainties, precision, in x and y from repeat measurements of each calibration standard. Uncertainties on both axes are homoscedastic,

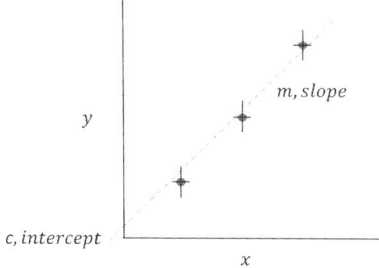

Figure 5.2 Linear regression including uncertainties in x and y values.

meaning they are the same magnitude for each concentration measurement point. It is not uncommon for the dependent y variable, the detector signal, to be heteroscedastic, where the uncertainty magnitude is a percentage of the signal output. Examples include optical pulses s^{-1} for measuring airborne particulate densities. In these instances, the calibration linear regression parameters can be adjusted using weighted least squares regression, so the higher uncertainties do not distort or overly influence the sample quantity estimates at lower levels.

Correct selection of an appropriate calibration range is vital for good quantitative measurement. Too broad a calibration range and the measurement precision, dispersion (uncertainty), may be too large. Ideally, calibration should be done in at least a five-point concentration range spanning at least two points above and two points below an expected sample concentration, and with a step change appropriate to the measurement resolution. While it may be considered ideal to have an intercept, c, value of zero, in reality there is always background signal noise coming from the instrument electronics and the sample matrix. It is therefore bad practice to force a 0,0 coordinate for x,y. This is making up data, and distorts the calibration linear regression resulting in a false gradient, m, value, introducing bias in the calibration gradient. If the calibration data is incorrect then most of the performance characteristic values will also be incorrect. Different types of calibration procedures are listed in Table 5.3.

Examples of non-linear calibration responses are pH and ion selective electrodes or wind speeds, bacteria, and pollen counts. Linear regression is the most common instrument detector response to amounts of substance. Some instruments use hardware and/or software for automatic linearization of detector signals and examples are semiconductor type gas sensors. Calibration outliers should be investigated, confirming whether this is showing real phenomena or operator error. Before calibration, the analytical scientist must ensure

Table 5.3 Types of calibration procedures.

Calibration type	Description	Limitations
External	Use of pure standards of the chemicals for calibration, often in solution with the accompanying solvents in the final sample preparation.	Susceptible to increased random uncertainty, poorer measurement precision, particularly if analysing small volumes of calibration samples.
Standard addition	Known concentrations or amounts of standards added to a sample matrix that potentially also contains the chemical of interest. Extrapolation of the calibration gives the amount of chemical in the original sample.	While good for quantification of low concentrations, an added standard may not be in the identical form of the chemical of interest in the sample matrix, therefore not responding identically to the detector.
Internal standard	Addition of a fixed amount of chemical, not identical but as similar as possible to the chemical of interest, so exhibiting the same or similar chemical characteristics hence responding similarly to the detector. An isotopically labelled chemical of interest is an example.	Careful addition of quantities of internal standard, otherwise different amounts added in calibration standards and sample will introduce measurement bias. Even with small variation.
Matrix matched	Fixed amounts of sample matrix are spiked with the pure chemical of interest over the required calibration range. Optimum for accounting for detector response to the sample matrix.	Sample matrix may have chemical components that interfere with the analyte signal, suppression or enhancement, or introduce a high background noise so reducing detector sensitivity to the analyte.

a detector signal response is specific to the analyte of interest and is selective. Selectivity is *'the extent to which the method can be used to determine particular analytes in mixtures or matrices without interferences from other components of similar behaviour'*.[13] As mentioned above, some industries prefer to use the term specificity.

5.5.2 Matrix Effects

Matrices are many and varied and so are matrix effects on a detector, which need explicit exploration in a method validation process. The International Union of Pure and Applied Science (IUPAC) regards a matrix effect as the *'combined effect of all components in a sample other than the analyte on the measurement of a quantity'*.[14] Matrix effects often increase instrument limits of detection and limits of quantification, thereby reducing instrument sensitivity to analytes of interest. Any specific chemical that causes a change in the measurand analyte detection is known as an 'interferent'. Physical properties such as temperature may also influence a detector signal output; examples are non-dispersive infrared, or in headspace analysis where the sample and headspace temperatures are important for method optimisation. Interferents can enhance as well as suppress detector signals, causing distortions in output data compared to that from pure standard chemicals. The optimum approach is to eliminate any impact of matrix components on a detector signal and data but this may be difficult, time consuming, and costly. Instead, be aware that adding pure standards may not be representative of the chemical form of a measurand of interest in the sample matrix. Sample preparation processes may replace one matrix effect with another. See examples given in earlier chapters. A good strategy is to check for any impact of matrix effects on all performance characteristics and to quantify and document these, referencing against measurements of analytes in simple and alternative matrices and controlling variables wherever possible.

5.5.3 Working Range

For the reasons mentioned in Section 5.5.1, the working range performance characteristic for quantification is too broad and it is unlikely that a method development needs an instrument detector to measure an analyte concentration as broad as the working range. The general definition of the 'working range' is the range over which a method gives results with an 'acceptable' uncertainty. Acceptable depends on measurement requirements and tends to require homoscedastic uncertainty across the measurement range minimum to maximum boundaries. The lower, minimum boundary of the range is the method limit of quantification (MLQ) or limit of quantification (LOQ). The upper, maximum boundary of the range is before any drop in the detector resolution, near saturation or due to other interference problems. Quantification method development should instead set and make use of an appropriate calibration range, as described in Section 5.5.1. Visual inspection of linearity across a working range can be misleading and its use can add bias and systematic uncertainty in quantification results. This is analogous to looking at a mountain range from a great distance, the heights look similar, but closer up, the variation in peak differences and gradients become increasingly apparent. A linear regression gradient over a narrower calibration range will be different to that over the entire working range. This is of particular concern close to quantification limits where these can become grossly distorted when using the working range.

5.5.4 Limits of Detection and Quantification

Limit of detection data is helpful when choosing an instrument. When selecting measurement methods by far the best approach is to avoid measuring anywhere near an instrument detection and quantification limit. Measuring at the limits of instrument sensitivity substantially increases the risk of false negative results. Limits of detection, a clear signal from a contaminant above the baseline signal, may however be useful for identifying low level contamination. A limit of detection may also be known as a 'critical value' or 'limit of a blank' or 'decision limit', each of which apply to a specific context, for example in analytical toxicology, clinical chemistry, and biopharmaceuticals.[15] It is important to know that the level of confidence in a detector signal is a signal and not just noise, or to have confidence that a predefined low-level limit is not exceeded. Other terms include 'critical concentration at risk $\alpha(CC\alpha)$', the decision limit, and 'critical concentration at detection risk β ($CC\beta$)', the detection limit.[16,17] These tend to be used for compliance tests, assessing risks of false positives or negatives rather than the actual instrument limit of detection for a given analyte. Limits of detection or quantification are also known as performance limits. Better terms are method limit of detection (MLD) and method limit of quantification (MLQ) as these should represent measurement limits associated with a given sample matrix and analytical method. The best types of samples for determining detection and quantification limits are 'blank matrices' where all chemical components, other than the chemical measurand of interest, are present. This approach needs the blank to give a measurable instrument signal, a measurable noise. If not, then very low-level spiking of concentrations near the limit of detection (LOD) and limit of quantification (LOQ) is necessary. The blank matrix is then subject to the entire documented analytical measurement process, sample preparation, clean up, pre-concentration, derivatisation, solution make up, addition of internal standards, and instrument analysis. Reagent blanks are more closely representative of an instrument limit of detection rather than a limit of detection from the entire analytical process. At least 6–10 replicate blank measurements give a blank measurement variance, where the 95th percentile of variance is typically used. Intermediate precision is a better approach to determining MLD and MLQ. It is advised to measure six replicate blanks, each day, over an extended period of time, under the identical procedure, using same instrument and set up conditions, by the same analyst, or analysts of equivalent competence.

The blank intermediate precision median or mean and standard deviation are given in eqn (5.2) and (5.3). The intermediate precision, s_0 (note, s_0^2 = blank sample variance and s_0 = blank sample standard deviation), is given by:

$$s_0 = \sqrt{\frac{1}{n-1} \sum_{i=1}^{n} (x_i - \bar{x})^2} \qquad (5.2)$$

After establishing s_0 the subsequent blank correction using a single blank measurement, the blank correction, s_0', is given by:

$$s_0' = s_0 \sqrt{\frac{1}{n_{\text{b.i.p.}}} + \frac{1}{n_{\text{blank}}}} \qquad (5.3)$$

where $n_{\text{b.i.p}}$ is the number of blank samples used to estimate the blank intermediate precision and n_{b} is typically a single blank run consecutively when also measuring the chemical of interest. The intermediate precision blank should not be used if blank values vary significantly over time, *i.e.* vary from day to day. Six blank measurements should ideally be run each time samples are measured, particularly if detection is likely to be near the limits, which, as stated above, is best avoided.

If no blank correction is needed then the standard error of the blank means, $s_{\hat{x}_0}$, eqn (5.4), might be used to estimate the MLD and MLQ, eqn (5.5) and (5.6). Be aware that this is considered to represent a stable distribution of randomly sampled means from a large replicate population.

No blank correction, $s_{\hat{x}_0}$:

$$s_{\hat{x}_0} = \frac{s_0}{\sqrt{n_{\text{b.i.p}}}} \qquad (5.4)$$

There are various ways to calculate the LOD and MLD, so it is important to state the method used in documentation. The simplest is the signal to noise ratio approach, $3s_0$, but this is susceptible to false positives and negatives. The International Union of Pure and Applied Science (IUPAC) method is statistically more robust.[18,19] This uses the standard deviation of 6–10 replicate blank samples (s_0) multiplied by a suitable factor, k_{limit}, which gives a low probability of type I (false positive) and type II (false negative) errors. k_{limit} is based on Student's t distribution, as repeat analytical blank measurements are reasonably assumed to follow a normal (Gaussian) distribution and the central limit theorem. The critical Student's t value for 10 000 plus replicate measurements, considered a population normal distribution, is $t_{critical}(0.05, \infty) = 1.65$. Using a one tail Student's t test, one tail reflects interest in only the distribution upper 95th percentile, therefore a significance, α, of 0.05 shows the one tail threshold limit of 95% of the replicate detector response to a blank. Infinite degrees of freedom, v, means 10 000 plus replicates in theory (try this in MS Excel or find in t tables). Recalling CCα and CCβ above, $\alpha = \beta = t_{critical} = 1.65$. Use of CC$\alpha$ and CCβ is a little ambiguous as strictly these are documented as compliance limits as opposed to parameters for estimating MLD or LOD despite using the same terms α for detectable concentration limit and β for detector signal limit. The scaling factor is therefore $k_{limit} = 1.65 + 1.65 = 3.3$, hence the IUPAC MLD and LOD are estimated using eqn (5.5).

Method limit of detection, MLD, or LOD:

$$\text{MLD or LOD} = \bar{x}_0 + 2(1.65) \times s_0 = \bar{x}_0 + 3.3 s_0 \quad (5.5)$$

where $\bar{x}_0$ is the mean blank concentration, in appropriate units, from replicate blank measurements. If a blank cannot be measured, then use a low-level concentration spike and associated linear regression to convert a blank signal output. The MLQ or LOQ does not use the Student t test statistical basis directly but follows this instead by setting an arbitrarily higher limit for confidence in quantity values. MLQ and LOQ can be estimated using eqn (5.6).

Method limit of quantification, MLQ, or LOQ:

$$\text{MLQ or LOQ} = \bar{x}_0 + 6(1.65) \times s_0 = \bar{x}_0 + 10 s_0 \quad (9.9 \text{ rounded up to } 10) \quad (5.6)$$

Uncertainty, dispersion – precision, s_0, and limit values of blank and low-level concentrations, are often considered unacceptably high and are susceptible to change, hence why it is best to avoid taking measurements near detection limits.

5.5.5 Sensitivity

Sensitivity is effectively the detector response to a given analyte, which is likely to be different for different analytes. A gas chromatography flame ionisation detector sensitivity reflects the ionizability of the chemical species arriving at the hydrogen and air flame. The molar absorptivity (extinction coefficient) following the Beer–Bouguer–Lambert law depends on the pi bonding delocalised electron chemical structure and sample matrix. Fluorescence emission can be highly susceptible to quenching from a sample matrix. Resolution, the size of a measurable step change in detector response with a step change in analyte concentration, $d[\text{detector signal}]/d[\text{analyte}]_{\text{matrix}}$, is a quantification of sensitivity. The smaller the step change, the more sensitive the detector is to the analyte. It follows that the smaller the detector signal is for a given analyte, the less responsive the detector, and the poorer the resolution, unless the signal is amplified. A correlation between detector response and analyte quantity may not necessarily be linear, in these instances the sensitivity may therefore change with the amount of analyte. It is not always the detector response that changes, rather it is the chemicals of interest that change, either deliberately, as in flame ionisation detection, or inadvertently, as with analyte photodegradation. Methylene blue is useful for monitoring reaction kinetics, where a change in methylene blue concentration occurs and the absorption detection resolution decreases with time. This is not a change in detector resolution response, but clearly what is happening chemically within the sample matrix.

5.5.6 Accuracy (Trueness) and Precision

The terms 'accuracy' and 'precision' are applicable to qualitative and quantitative analysis. For qualitative analysis, accuracy might refer to chromatographic retention time and relative retention times. Retention time standard deviation from repeat analysis indicates a qualitative precision. In mass spectrometry, mass to charge ratio (m/z) values show the accuracy of identification relative to instrument tuning and pure standard compound expected m/z fragment values, and also repeatable m/z values to a given number of significant figures, hence accurate mass. Trueness is accuracy if referring to a replicate measurement population, a very large number, of replicate analysis referring to an *expected* central parameter value, typically mean or median. Accuracy and 'trueness' refer to a known true value that is established in the wider scientific field such as melting points or boiling points of pure chemicals or matching a measurand value to that from a reliable certified reference material (CRM). Precision is how close each measurement, in a set of replicate measurements, is to the accurate, or known true value. Precision is more than accuracy as it is a dispersion parameter. As discussed above in Section 5.5.4, when defining method detection and quantification limits it is reasonable to assume that replicate measurements of assumed identical, or near identical samples, independent of one another, follow a normal (Gaussian) distribution. Each unique normal distribution is described by a mean or median central value and variance (standard deviation) dispersion of measurement values either side of the central value, $\bar{x} \pm s_x$. A population distribution can take on two seemingly contradictory definitions.

A. Central limit theorem: After 10 000 or more replicates there is little to no change in the central value, mean or median, and the distribution becomes more leptokurtic. More of the repeat values are closer to the central value. Assuming an estimation of the standard error of the mean, σ_μ applies, then the standard deviation of a set of mean values from several large sets of replicate analysis, $\sigma_\mu = \dfrac{\sigma}{\sqrt{n}}$, gives a quantification of how well a mean represents the population mean.

B. A three sigma rule, 3σ, gives a not unreasonable assumption that for 10 000 or more replicates randomly drawn from the same population, each replicate being independent of the other, will result in no greater dispersion than $\pm 3\sigma$. The central limit theorem also applies as after 10 000 or more replicates there is little to no change in the central value, mean or median value.

A. indicates that, with a greater number of replicate samples there is increasing likelihood of more replicate measurements being closer to the central mean value, and the spread of values, dispersion, getting smaller. This gives a more leptokurtic population distribution than a distribution with fewer samples. B. indicates the intuitive and often real measurement data experience as with an increasing number of replicates there is still an increasing likelihood of more replicate values being close to the central mean value. However, there is also a likelihood of greater dispersion, up to a maximum of the ± 3 sample sigma limit, for 99.7% of replicate samples, with both A and B having a 0.3% asymptote at the lower and upper ends of the normal distribution – see Figure 5.3.

There is much crossover in the use of accuracy and precision, particularly when various authorities insist that their definition is correct without including statistical clarification. The simplest approach is to consider accuracy as a measurement value degree of match to a target value, either a single value, or better still, the central

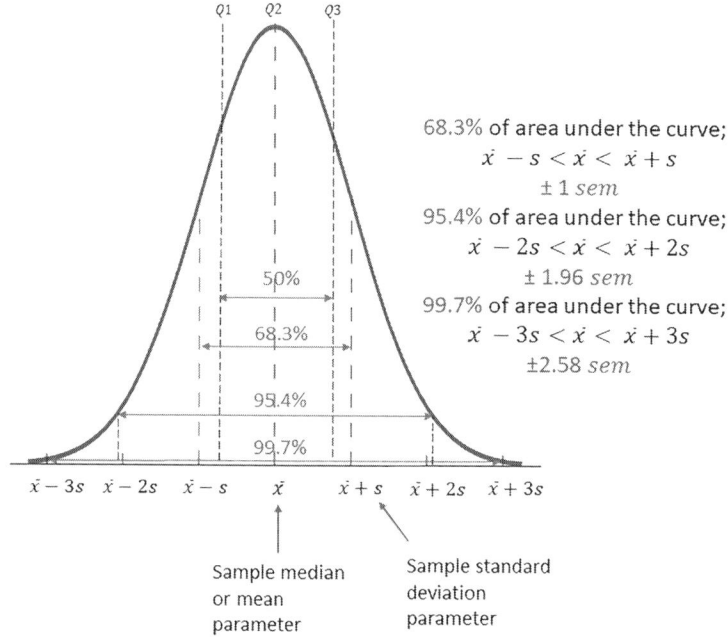

Figure 5.3 A normal (Gaussian) distribution, probability density function, of randomly drawn replicate sample amounts, x, independent of one another. The asymptote at the x axis when f(x) is close to zero width is exaggerated for illustration purposes.

mean or median value from replicate measurements. Caution must be exercised to not introduce bias when using a target value and accepting a value as close to a target when in reality it might not be, hence with reference to validity and reliability checks. Precision is the likelihood of achieving the target value, so can be quantified as replicate measurement dispersion around the target value – variance (standard deviation), standard error of the mean. Precision, measurement replication, is one of the most important and essential method performance characteristics. Ensuring representativeness of replicate samples, includes the sample collection, preparation and ideally measurements that are matrix matched. As already noted, precision terms are repeatability, intermediate precision, and reproducibility. Reproducibility may be useful for reflecting impact from small changes in sample preparation techniques and different types of measurement instrumentation and is often used in inter laboratory comparison. Performance characteristics can include precision limits. Following the statistical methods for precision in method limits of detection in Section 5.5.4, (i) a repeatability limit uses for example at least six repeats instead of a population of repeats and (ii) uses a two tail sample normal (Gaussian) distribution rather than a one tail limit threshold, as in eqn (5.7).

$$\text{repeatability} = \sqrt{6} \times t \times s_r \qquad (5.7)$$

where s_r is the standard deviation of six repeat measurements, or intermediate precision, s_I or reproducibility, s_R. Repeatability of the sample set may also be simply represented as s_r. Some practitioners assume two replicates randomly drawn from a population is sufficiently representative, this is potentially highly speculative unless providing solid and reliable evidence to support this assumption. Inferential statistics such as single factor analysis of variance (ANOVA), Table 5.4, can be used to assess intermediate precision and reproducibility, eqn (5.8a) and (5.8b):

$$s_I = \sqrt{s_r^2 + s_b^2} \qquad (5.8a)$$

where s_I is the intermediate precision, s_r^2 is the repeatability variance, s_r is the repeatability standard deviation, also in eqn (5.8b). MS_w is the within sample mean square estimate and s_b^2 is the between sample (between group factor) variance, see Table 5.4

$$s_r = \sqrt{MS_w} \qquad (5.8b)$$

The single factor is the different methods, or different analysts, or different repeat sets of data. It is not advisable to state time such as days as the single factor. Time is a masquerading factor as this is irrelevant to the measurement output data compared to other variables. It is not usually the time that might change a measurement output. Two-factor ANOVA could be used to compare intermediate precision with reproducibility.[20]

Table 5.4 Single factor analysis of variance (ANOVA).

Source of variation	Sum of squares	dof, ν	Mean square	F ratio
Between groups	$\sum_{i=1}^{k} n_i(\bar{y}_i - \bar{Y})^2$	$k-1$	$MS_b = s_b^2$ $= SS_b/k-1$	$\dfrac{MS_b}{MS_w}$
Within groups	$\sum_{i=1}^{k}\sum_{j=1}^{N}(y_{ij} - \bar{y}_i)^2$	$nk-k$	$MS_w = s_w^2$ $= SS_w/n-k$	
Total	$\sum_{i=1}^{k}\sum_{j=1}^{N}(y_{ij} - \bar{Y}) = SS_B + SS_W$	$nk-1$		

5.5.7 Bias

Bias, also known as systematic uncertainty, is the deviation from an *expected* (accurate or true) quantity value. This is an agreed, established, reference measurement data value. Bias, systematic uncertainty, is difficult to identify, but once identified it is usually easier to correct for than the random uncertainty, the precision. Running repeat measurements to ensure a consistent bias rather than simply being an artifact of expected random variation, ensures precision. Reference measurement values are from reference materials, the best of which are certified reference materials (CRMs), and/or recoveries from spiked sample matrices or comparing with results established elsewhere, using more than one method, comparing measurement data from a new method to a reference method, or analysis by different laboratories. These require further performance characteristic checks, considering levels of precision before determining any bias. A two-dimension plot of calibration and measurement output from two methods is a good way to visualise any differences between the methods. A gradient of 1 and intercept of zero indicate the data outputs from both methods are identical. This does however need multiple data points and precision in concentration and the signal output needs to be small and homoscedastic. Estimation and plotting confidence intervals of 95% or 99% to support interpretation of significance in differences or not, and a plot of the residuals is also a useful indication. A significant contribution to measurement bias may come from incorrect calibration – calibration bias. Higher or lower quantities of external standards for each calibration concentration, or a more or less internal standard than thought, introduces calibration bias. When using reference materials for method validation the same reference material should not be simultaneously used for bias checks and matrix matched calibration as any bias is carried through, so distorting the entire calibration. There are many potential contributions to systematic uncertainty as discussed, with examples, in the previous chapters. Quantifying systematic uncertainty, bias, can simply be a difference between the *average observed* and *expected* values, eqn (5.9), or relative to a reference material, a CRM, eqn (5.10), or recovery from a spiked matrix, eqn (5.11), or just a relative recovery, eqn (5.12).

$$u_{systematic} = \bar{x} - x_{reference} \tag{5.9}$$

$$u_{systematic}\% = \frac{\bar{x} - x_{reference}}{x_{reference}} \times 100 \tag{5.10}$$

$$u_{systematic}\% = \frac{\bar{x} - x_{spike}}{x_{spike}} \times 100 \tag{5.11}$$

$$\text{Recovery \%} = \frac{\bar{x}}{x_{ref \text{ or spike}}} \times 100 \tag{5.12}$$

Eqn (5.7)–(5.12) are simple routine checks typically used to ensure the method performance is as expected. Interpretation is important as sample preparation and extraction recovery is the extraction not the measurement bias. Measurement bias is any deviation from expected recovery quantities. Poor recoveries mean improvements to the extraction methods needs consideration, particularly for measurands that are inherent in sample matrices, rather than the less representative spiking of sample matrices. Repeat analysis of a reference material gives a median and standard deviation of $\bar{x}_{reference} \pm s_{reference}$, but reference materials can be expensive, so this may only be done during method development and initial method set up and periodically rather than routinely. When developing a method, significance of any systematic uncertainty, bias, should be checked using inferential statistics of Student t tests or analysis of variance (ANOVA). Where there is indication of any physical and or chemical variable impacting measurement output data, hence causing bias, tests such as analysis of covariance (ANCOVA) should also be done during method development.[20] Table 5.5 and eqn (5.13) and (5.14) show estimates of any significant bias in measurement data from the same sample using two different analytical methods.

$$s_c = \sqrt{\frac{0.4^2 \times (6-1) + 0.3^2 \times (6-1)}{6+6-2}} = 0.354 \tag{5.13}$$

$$t = \frac{(2.75 - 2.6)}{0.354\sqrt{\frac{1}{6} + \frac{1}{6}}} = \frac{0.15}{0.204} = 0.735 \quad \text{this is the } t \text{ estimate} \tag{5.14}$$

Table 5.5 Measurement of mean and standard deviation from six replicate measurements from using each of two different analytical methods.

	$\bar{x}$ µg cm^{-3}	s µg cm^{-3}	n
Method 1	2.75	0.4	6
Method 2	2.6	0.3	6

where s_c is the pooled standard deviation. The t test should be run with no significant difference between the factor level variances, but a pooled t test selects a reasonably representative worst case variance for the given data.

$t_{critical}\,(\alpha_{0.5},\,\upsilon_{10}) = 2.23$ at significance $\alpha = 0.05$ and $\upsilon = 10$ degrees of freedom. $t_{critical} > t_{estimate}$, inferring there is no significant difference between the means of the two methods. The difference between the means is 0.15, from eqn (5.14), *i.e.* it is clear that the adjusted 'difference' standard deviation of 0.204 is greater than 0.15, hence giving a measurement bias uncertainty of 0.204. If the measurement bias increases over time, *i.e.* it drifts away from the expected performance characteristic values previously recorded, then this indicates a lack of ruggedness or robustness in the measurement method.

5.5.8 Ruggedness, Robustness

A distinction between ruggedness and robustness is given in the introduction to this chapter. These terms can be used interchangeably. Ruggedness is the resilience of a method's output variables (measurement data) to expected small changes in input variables, such as temperature, volume, pH, and so on, often within the analytical scientist's control. Robustness is resilience to changes in external variables, which can also be temperature, volume, pH, and so on, but where the analytical scientist is less able to control these. Testing for ruggedness and robustness can give an indication of the degree of care needed when carrying out a measurement method and particular techniques, and guide experiment design modification. Every step in the analytical method, from sampling to storage and to instrument measurement, needs careful scrutiny of the method for ruggedness and robustness. Collection and storage of samples in polyethylene terephthalate can result in contamination of the sample with phthalate ester plasticisers. Changes in sample temperature during storage can result in water moisture condensation causing degradation or reactivity. Testing method ruggedness and robustness at the simplest level is effectively method stability testing. A more complex experiment design evaluation that considers multiple factors and variables uses chemometrics tools such as factorial, 'level 2', 'Plackett–Burman' screening, or 'response surface', multivariate analysis, and so on.[20,21] The measurement method design choice depends on available resources and the degree of control needed. The trick is to identify which variables pose the greatest threat to the method's robustness and ruggedness and hence those that increase

Table 5.6 Selection and scaling variables, 'two-level' design.

	Factor 1	Factor 2
Trial 1	−1	−1
Trial 2	+1	−1
Trial 3	−1	+1
Trial 4	+1	+1

random and/or systematic uncertainty, then changing the variables one at a time, particularly if more than one variable is identified as impacting robustness. Consider starting with a 'level 2' design, changing variables at low and high levels, and fixing any other factor variables as necessary – see Table 5.6. An additional column can be added if three factors need consideration. Variables can then be ranked in order of impact on method performance, followed by statistical tests to indicate the significance of the observed impact. These design tests evidence a need to control any variables in the method and to what degree. Method controls and consequences to measurement data output should be clearly documented in method procedures during method development, and these should also be noted clearly in publications. Unfortunately, such method design details can be lacking or missing in published methods.

5.6 Measurement Uncertainty

A definition of uncertainty in analytical measurement is *parameters characterising a measurement dispersion and measurement difference*. Exploring and understanding the causes of *dispersion* around a central, target measurement value – *random uncertainty*, and deviation, *difference*, bias from a central target value – *systematic uncertainty*. Understandably, uncertainty is often interpreted as doubt in the measurement data but should never be of thought of in this way. Instead, consider uncertainty as an inextricable parameter of the measurement data, rich with useful information on method performance. Knowing the uncertainty parameters of a measurement increases the confidence in, and the validity of, a measurement result. Sources of uncertainty are many and varied and examples are covered extensively in the previous chapters. Errors and uncertainties are referred to interchangeably. Error is considered to be (i) the difference between a target reference (expected) and measured (observed) value and (ii) used when the target is an established, true value and is widely accepted and documented, such as the boiling point of a pure liquid. When the 'true' value is unknown until after measurement, the unbiased parameter estimate should be referred to as the measurement uncertainty. Error is quoted as a single value, whereas uncertainty is typically a range or distribution, particularly the random uncertainty as in precision estimates described in Section 5.5.6. The ISO/EC validation in measurement (VIM) definition of uncertainty is '*a non-negative parameter characterising the dispersion of the quantity values being attributed to a measurand, based on the information used*'.[22] Again, as described earlier in Section 5.2, 'measurement uncertainty from sampling' can be determined empirically through measurement performance characteristics, and/or using model propagation of predetermined contributions to the overall measurement uncertainty. An important and potentially significant contribution to the measurement uncertainty budget might be from the data processing methods and use of models. Inappropriate model selection or truncating data values, using too few significant figures in calculations, can distort data output. Validity and reliability of measurement results must therefore include uncertainty traceability.

5.6.1 Uncertainty Traceability

Traceability relies on accurate and precise documentation, the chain of sample transfer, preparation, matrices, stability, environment, and analysis, including all calculations and data processing. Any contribution to the uncertainty budget should be documented throughout the analytical process, using pre-defined standards, weighing balance and equipment calibration, and ensuring consistency such as use of SI units. This guides a consistent standard of practice within and between laboratories and *in situ* measurements. Descriptions of activities for ensuring traceability in chemical measurement processes are in the Eurachem CITAC Guide 'Traceability in Chemical Measurement'.[23] These include measurand specification and target measurement uncertainty, using selection and careful documentation of the measurement procedure and selection of appropriate standards of measurement. Measurand specification includes analyte specificity, physical form, mass fractions, isomers, and isotopes. It is crucial to have confirmation of the source and identity of reference materials for true traceability and the uncertainties associated with measurements of these reference materials. For the sample measurand, uncertainty definition is relative to uncertainties in reference

Validity and Reliability of Analytical Measurement Methods 73

measurements. Clear definition of contributions to uncertainty, and methods used for estimating these, must be traceable.

5.6.2 Estimating Uncertainty

Individual and overall precision and bias uncertainty estimation are given in Sections 5.5.6 and 5.5.7. After clear specification of the measurand and associated analytes, it is important to identify and quantify uncertainty sources. As already highlighted, sampling is often a significant source of uncertainty. Storage, sample preparation, reagent impurities, matrix effects, reagent purity, reaction stoichiometry unknown or assumed, operator care and mistakes, environment and measurement conditions, are all contributions to random uncertainty (at typically every step in a process) and data processing calculations. If a process from where the sample comes from, for example food products from cereals, is long established then prior sample population data from a data bank could be a good reference for uncertainty quantification. The best available estimate of overall precision and best available estimate of overall bias is often sufficient for in house method development and validation. Any changes to robustness or ruggedness must also have accountability in uncertainty budgets. To guide method performance improvement,

Table 5.7 Combined uncertainty modelling.

Function	Formula
$z = a+b$	$\delta z = \sqrt{(\delta a)^2 + (\delta b)^2}$
$z = ab$ $z = a/b$	$\left(\dfrac{\delta z}{z}\right)^2 = \left(\dfrac{\delta a}{a}\right)^2 + \left(\dfrac{\delta b}{b}\right)^2 \quad \delta z = z\sqrt{\left(\dfrac{\delta a}{a}\right)^2 + \left(\dfrac{\delta b}{b}\right)^2}$
$z = \dfrac{ab}{cd}$	$\delta z = z\sqrt{\left(\dfrac{\delta a}{a}\right)^2 + \left(\dfrac{\delta b}{b}\right)^2 + \left(\dfrac{\delta c}{c}\right)^2 + \left(\dfrac{\delta d}{d}\right)^2}$
$z = \dfrac{a^n b^m}{c^p d^q}$	$\delta z = z\sqrt{\left(n\dfrac{\delta a}{a}\right)^2 + \left(m\dfrac{\delta b}{b}\right)^2 + \left(p\dfrac{\delta c}{c}\right)^2 + \left(q\dfrac{\delta d}{d}\right)^2}$
$z = e^x$	$\delta z = e^x \delta x$
$z = \ln x$	$\delta z = \dfrac{\delta x}{x}$
$z = \dfrac{1}{x}$	$\delta z = \dfrac{\delta x}{x^2}$

uncertainty budget modelling will help to identify the largest contribution to uncertainty, assuming all contributions are known. See earlier discussions in this chapter on using uncertainty modelling. Table 5.7 lists uncertainty models and assumes that each contribution is independent of other contributions, *i.e.* there are no correlations. The combined uncertainty model in the first row shows propagation of absolute uncertainties, hence with common units, and the second row is the propagation of relative uncertainties, which are non-identical properties so are dimensionless.

References

1. ISO/IEC 15189:2022, *Medical Laboratory Requirement for Quality and Competence*, Ed.4, reviewed and confirmed 2023. Available from: https://www.iso.org/standard/76677.html [accessed 28th May 2025].
2. ISO/IEC 17025:2017, *General Requirements for* the Competence and Testing and Calibration Laboratories, Ed.3, reviewed and confirmed 2023. Available from: https://www.iso.org/standard/66912.html [accessed 28th May 2025].
3. M. Sargent, *Anal. Proc. incl. Anal. Commun.*, 1995, **32**(5), 201–202.
4. *The Fitness for Purpose of Analytical Methods – A Laboratory Guide to Method Validation and Related Topics*, ed. H. Cantwell, Eurachem Guide, 3rd edn, 2025. Available from https://www.eurachem.org/index.php/publications/guides/mv [accessed 28th May 2025].
5. M. Thompson, S. L. R. Ellison and R. Wood, *Pure Appl. Chem.*, 2002, **74**(5), 835–855.
6. P. M. Gy, *Sampling of Heterogeneous and Dynamic Material Systems*, Elsevier, Amsterdam, 1992, E Book ISBN: 9780080868370. Available from https://shop.elsevier.com/books/sampling-of-heterogeneous-and-dynamic-material-systems/gy/978-0-444-89601-8.
7. *Measurement uncertainty arising from sampling – a guide to methods and approaches*, ed. M. H. Ramsay, S L. R. Ellison and P. Rostron, Eurachem Guide, 2nd edn, 2019, ISBN: 978-0-948926-35-8. http://www.Eurachem.org
8. ISO/IEC 9000:2015, *Quality management systems, fundamentals and vocabulary*, Ed.4, reviewed and confirmed 2024. Available from: https://www.iso.org/standard/45481.html [accessed 29th May 2025].
9. A. N. Analytical Methods Committee, *Anal. Methods*, 2015, **7**(18), 7404–7405.
10. ASTM E1601-19, *Standard practice for conducting an interlaboratory study to evaluate the performance of an analytical method*, 2019. www.astm.org.
11. CEN/TR 10345:2013, *Guideline for statistical data treatment of inter laboratory tests for validation of analytical methods*, CEN Brussels.
12. ISO/IEC 17034:2016, *General requirements for the competence of reference material producers*, Ed.1, reviewed and confirmed 2022. Available from: https://www.iso.org/standard/29357.html [accessed 3rd June 2025].
13. J. Vessman, R. I. Stefan, J. F. Van Staden, K. Danzer, W. Lindner, D. T. Burns, A. Fajgelj and H. Muller, *Pure Appl. Chem.*, 2001, **73**(8), 1381.
14. A. D. McNaught and A. Wilkinson, *Compendium of chemical terminology*, 2nd edn, (the 'Gold Book'), Compiled by Blackwell Scientific Publications, Oxford, 1997, Online version (2019) created by S. J. Chalk, ISBN 0-9678550-9-8. https://goldbook.iupac.org/ [accessed 12th June 2025].
15. T. A. Little, *BioPharm Int.*, 2015, **28**,(4), 48–51.

16. Commission Regulation (EC) No 657/2002, 12 August 2002, implementing Council Directive 96/23/EC concerning the performance of analytical methods and the interpretation of results, Of. J. EU, L 221, 17 August 2002. [Accessed 12th June 2025].
17. J. Van Loco, A. Jànosi, S. Impens, S. Fraselle, V. Cornet and J. M. Degroodt, *Anal. Chim. Acta*, 2007, **586**(1), 8–12.
18. L. A. Currie, Nomenclature in evaluation of analytical methods, including detection and quantification capabilities (IUPAC Recommendations 1995), *Pure Appl. Chem.*, 1995, **67**, 1699.
19. F. Allegrini and A. C. Olivieri, *Anal. Chem.*, 2014, **86**(15), 7858–7866.
20. M. James and N. Miller, *Statistics and Chemometrics for Analytical Chemistry*, 7th edn, Pearson Education, 2020.
21. National Institute of Standards and Technology (NIST), *Engineering and Statistics Handbook, 5.3 Choosing an Experiment Design*. https://www.itl.nist.gov/div898/handbook/pri/section3/pri3.htm [accessed 24th June 2025].
22. ISO/IEC 99:2007, *International vocabulary of metrology – Basic and general concepts and associated terms*. https://www.iso.org/standard/45324.html [accessed 24th June 2025].
23. Eurachem CITAC Guide, *Metrological Traceability in Chemical Measurement*, 2019. https://www.eurachem.org/index.php/publications/guides/trc [accessed 27th June 2025].

Subject Index

accelerants, 20
accuracy, 4, 39, 48, 64–65, 71
acoustic emission analysis, 2
ACS Omega, 14
adenosine triphosphate (ATP), 25
Aflatoxin B1, 5, 7
agglomeration, 50
aggregation, 27
ALCOA (attributable, contemporaneous, original, accurate), 41
amphetamines, 35
analogues, 35
analysis of covariance (ANCOVA), 69
analysis of variance (ANOVA), 67, 69
analyte recovery, 6
arsenic, 2
atomic absorption spectroscopy (AAS), 33

Benchmark (digital platform), 12
bias, 2, 7–8, 27, 42, 50, 53–55, 57–60, 68–71; *see also* uncertainty
bicarbonate, 8, 24
biosensors, 14–15
blanks, procedural, 52–53
blank samples, 40, 42, 52, 61–63
blood, 23, 29, 36, 39–40
boiling points, 64, 72
bulk materials, 28–29, 42, 49–50

caffeine, 34
calcium carbonate, 24
calibration, 20, 48, 52–53, 55–56, 56–59, 68
    concentration values, 57, 68
    outliers, 58
    reproducibility, 57
    standards, 3, 47, 53, 59
calibration bias, 68
carbon, inorganic, 8, 24
carbon dioxide, 8, 12–13, 24, 26, 29, 38
carbonic acid, 8, 24
central limit theorem, 56, 63, 65
certified reference materials (CRM), 8, 53–54, 57, 64, 68
characterisation, 19–20, 22
chemiluminescent spectroscopy, 33
chromatography, 3, 20, 32–33, 35
coccolithophores, 24
contaminants, 20, 34–35, 38–42, 50, 53, 61, 70

Control of Substances Hazardous to Health (COSHH), 15
CRM, *see* certified reference material
cyanuric acid, 42

data, specification and quality, 7–9
data processing, 4, 7, 22, 38, 48–49, 54, 72
data requirements, 21, 31, 49, 52
decision limit, *see* limits of detection
derivatisation, 24, 35, 52–53, 61
detection limits, 37, 48, 61, 63
detectors, 2, 29, 31–35, 55, 59, 64
dimethyl phthalate, 34
dioxins, 21
disintegration, 50
documentation, 48, 51–52, 54, 62, 72
drugs, 1, 7, 20–21, 23, 36

electron capture, 33
electron ionisation (EI), 35
electrospray ionisation (ESI), 6, 34
energy consumption, 11, 13, 35
environmental quality standards (EQS), 37
enzyme linked immunosorbent assay (ELISA), 43
ethanol, 25–26
Eurachem guides, 3, 8, 48, 50–51, 72
exploratory analysis, 19
extraction, 6, 24, 32, 34, 36, 38, 53, 69

false negatives, 52, 63
false positives, 7, 34, 37, 52, 61–63
fentanyl, 37
fermentation, 25–26
fishbone cause and effect diagrams, 2–3
fitness for purpose, 4, 19, 48, 51–52, 54
fluorescence spectroscopy, 20, 33
food samples, 7, 22, 39, 42, 47, 54
forensic science, 2, 20, 40, 47
freezers, 11, 13, 40
fundamental analysis, 22

gas chromatography, 10, 32, 35, 43, 63
gas sampling, 29
grain, 5, 28
Green Gown award, 12

Subject Index 77

heteroscedasticity, 58
homologous series, 35
homoscedasticity, 57, 60, 68
Horlicks, 1

inductively coupled plasma mass spectrometry (ICP-MS), 2
in situ sensing, 2, 14, 37–39, 49, 72
Institute of Occupational Safety and Health (IOSH), 15
instrument calibration, *see* calibration
instrument selection, 31, 32–33, 55
    limits of detection and quantification, 61
    performance characteristics, 33–34, 55–56
        selectivity, 34–35
        sensitivity and resolution, 36–37
    in situ sensing, 37–38
interlaboratory comparisons, 52
intermediate precision, 42, 62, 66–67
International Equipment Trading Ltd, 12
International Organisation for Standardisation/International Electrotechnical Commission (ISO/IEC), 3, 47
International Union of Pure and Applied Chemistry (IUPAC), 49, 59, 62
ISO/IEC 17025, 3
isomers, 19, 29, 35, 72

journals, 3, 13–14

Laboratory Efficiency Assessment Framework (LEAF), 11
laboratory information management systems (LIMSs), 41
lasers, 20, 22–23
LDPE (low density polyethylene), 39
Le Chatelier's principle, 24
limits of detection (LOD), 32, 36–37, 44, 61–63
limits of quantification (LOQ), 36, 38, 42, 44, 60–61, 63
linearity, 48, 56, 60
liquid chromatography, 6, 20, 32, 36, 42–43
liquid–liquid extraction (LLE), 42, 44
literature reviews, 3

mass spectrometry, 2, 5–7, 20, 33–37, 42–43, 64
    inductively coupled plasma (ICP-MS), 2, 36

ion chromatography tandem (IC-MSMS), 37
ion trap, 6
liquid chromatography tandem (LC-MSMS), 36
plasma based dielectric barrier discharge ionisation, 36
quadrupole, 35
triple quadrupole tandem, 6
matrix effects, 7, 33, 36, 48–49, 59–60, 73
    complex matrices, 1, 32, 36
measurand, 4–6, 19–20, 27–29, 32–33, 40–42, 52–54, 69, 72–73
measurement uncertainty, *see* uncertainty
measurement validity, *see* validity
melamine, 42, 45
melting points, 64
method development, 21–22, 24–25, 51, 54, 57, 60, 69, 71
    analysis types, 21–22
    performance priorities, 41–45
    sample preparation and collection, 27–29
    system responses, 26–27
method limit of detection (MLD), 36, 44, 61–63
method limit of quantification (MLQ), 36, 42, 44, 60–63
method performance characteristics, 32, 48, 55, 66
method validation, 52
methylene blue, 27
microfluid analysis, 20
mixtures, 27, 35
molecular spectroscopy, 21

negative results, 13–14, 61
non-destructive analysis, 20
nuclear magnetic resonance (NMR) spectroscopy, 5, 7, 33, 36

ochratoxins, 28

perfluorooctanoyl chloride, 35
personal protective equipment (PPE), 17
pH, 8, 24–25, 36, 41, 48, 58, 70
phase separation, 21
photoionisation detectors (PIDs), 25
photon emission spectroscopy, 33
poisons, 1, 20
population distributions, 56, 65

precision, 4, 8, 23, 37–39, 42, 53–59, 62–64, 66–68, 72–73
    accuracy and, 65
procedural blanks, 52–53

qualitative analysis, 19, 22, 44, 52, 64
quality control, 8, 25, 48, 52
    wine production, 25
quantification limits, 38, 48, 55, 61, 65
quantitative analysis, 19–20, 22, 44, 57, 64

radiality analysis of single puncta (RASP), 37
Raman spectroscopy, 33
reference materials, 20, 53–54, 68–69, 72
registration, evaluation, authorisation and restriction (REACH), 13, 15
reliability, 15, 21–23, 26, 45, 47–48, 51, 66, 72
repeatability, 8, 23, 29, 50, 57, 66–69
replicates, 56–57, 63–67, 69
risk assessment, 10–11, 15, 17, 52
robustness, 48, 55, 70, 73
ruggedness, 8, 48, 70, 73

safe limit values, 42
safety, 5, 8–10, 14, 16–17, 19, 51–52
    risk assessment and, 15–17
sample collection, 38–40
sample inlets, 33
sample matrix, see matrix effects
sample preparation, 6, 14–15, 27–29, 33–34, 40–42, 44, 49–50, 55
sample referencing, 19–20, 60
samples
    continuity, 40–41
    reference, 19
sample sizes, 39, 42, 52, 55
sampling, 4, 28–29, 39–41, 50, 52, 54–55, 70, 72–73
    uncertainty, 49–50
        estimation, 49
satellite electromagnetic sensing, 2
screening, 19, 21–22
selectivity, 4, 7–8, 23, 33–36, 42, 44, 48, 59

semi-quantitative analysis, 19, 21–22
sensitivity, 7, 23, 37–38, 63–64
soil environments, 1, 8, 13, 24, 36, 38
solid phase extraction (SPE), 31, 44
solid phase micro extraction (SPME), 44
solvents, 10–11, 13–14, 20, 34, 53, 59
    flammable, 11
    hydrocarbon, 10–11, 13–14
Student t test, 69
styrene divinylbenzene (SDB), 39
sustainability, 9–10, 15
    method development, 12–13
    negative results and, 13–15
    systems and networks, 11–12
Sustainable Laboratories Report, 11
system responses, 26–27

thermal conductivity, 33
thermal desorption spectroscopy, 33
three sigma rule, 65
traceability, 72–73
trace analysis, 7, 19–20, 33, 36, 39
tributyl phosphate, 29
trueness, see accuracy

uncertainty, 2–6, 8, 26, 45, 48–50, 54–58, 60, 71–74
    estimation, 8, 73–74
    systematic, see bias
    traceability, 72–73
uncontrolled parameters, 48
United Kingdom Accreditation Service (UKAS), 3

validation
    instrument calibration, 56–58
    matrix effects, 59–60
    planning, 54–55
    tools, 52–53
    validation samples, 52–54
validation in analytical measurement (VAM) principles, 47–48
validity, 7, 15, 23, 47, 51, 66, 71–72

Warp-it, 12
waste, 5, 10–15, 21, 36, 39
working range, 53, 60–61